T0261521

PTOLEMY'S

# *Geography*

J. Lennart Berggren

AND

Alexander Jones

# PTOLEMY'S

# *Geography*

AN ANNOTATED TRANSLATION

OF THE THEORETICAL CHAPTERS

PRINCETON UNIVERSITY PRESS

PRINCETON AND OXFORD

Published by Princeton University Press, 41 William Street,
Princeton, New Jersey 08540
In the United Kingdom: Princeton University Press,
3 Market Place, Woodstock, Oxfordshire OX20 1SY



ISBN 978-0-691-09259-1
This book has been composed in New Century Schoolbook.

The paper used in this publication meets the minimum requirements of ANSI/NISO Z39.48-1992 (R1997) (*Permanence of Paper*)

www.pup.princeton.edu

Printed in the United States of America
10 9 8 7 6 5 4 3 2 1

TO OUR TEACHERS

*Asger Aaboe*

*and*

*Gerald Toomer*

# *Contents*

# *Illustrations*

*Figures*



*Plates (following p. 128)*

*Maps*

# Preface

On any list of ancient scientific works, Ptolemy's *Geography* will occupy a distinguished place, and Ptolemy's contributions to ancient, medieval, and Renaissance geography receive respectful notice in books devoted to the history of science.

It is, however, the respect usually reserved for the dead, and one who wishes to examine the sources firsthand will soon come to an impasse. No complete edition of the Greek text has appeared since C.F.A. Nobbe's of 1843–1845; and while good translations of parts of the *Geography* have been produced in German and French, the only version in English has been the nearly complete, but in all other respects very unsatisfactory, translation by E. L. Stevenson.

The centerpiece of Ptolemy's book is an enormous list of place names and coordinates that were intended to provide the basis for drawing maps of the world and its principal regions. A reliable translation of this part, and of another long section consisting of descriptions (or, as we prefer to call them, captions) of the regional maps, is unattainable in our present defective state of knowledge of the manuscript tradition of the *Geography*. These passages are in any case not designed for continuous reading, and we believe that most readers will not be disappointed to find them represented here by a description and a representative excerpt.

The remaining chapters of the *Geography*, which we have translated in their entirety, may be read as a series of essays of varying length dealing with aspects of scientific cartography. The work has, however, a unity of purpose and design that is possibly easier to grasp when these theoretical sections are not overshadowed by the geographical catalogue.

In our introduction and notes, we have assumed that our likely readers will be varied: historians of science (and their students) wanting familiarity with a classic of science; historians of the ancient world, hitherto lacking convenient access to the work of its greatest mathematical geographer; and, finally, those geographers who are interested in the origins of their discipline. We are well aware, however, that the line between pleasing and offending these disparate groups is a thin one, and we can only take comfort in the closing words of Samuel Johnson's introduction to his edition of Shakespeare:

> It is impossible for an expositor not to write too little for some and too much for others. He can only judge what is necessary by his own experience: and how longsoever he may deliberate, will at last explain too many lines which the learned will think impossible to be mistaken, and omit many for which the ignorant will want his help. These are censures merely relative, and must be quietly endured.

This book had its beginnings at a time when one of us (AJ) was a student in the History of Mathematics Department at Brown University and the other (JLB) was there for a short visit. In the intervening years, the book took shape as each of us tested his contributions against the friendly skepticism of his co-author. In addition, a number of individuals and institutions have helped over this time, and it is a pleasure to thank them here: Gerald Toomer, for suggesting the project and encouraging the authors to believe that they could handle it; Asger Aaboe, who, when he heard of the project, gave JLB his copy of Nobbe's text of the work; Sarah Pothecary, for critically reading our early drafts; the Biblioteca Apostolica Vaticana, the British Library, the Bodleian Library, the Houghton Library (Harvard University), the John Hay Library (Brown University), and the Newberry Library, for supplying microfilms and photographs of manuscripts; the Natural Sciences and Engineering Research Council of Canada, for grants that made it possible for JLB to visit Brown University and to obtain photocopies of a number of rare printed editions of the Greek text of Ptolemy's work; our home institutions, for unfailing support of our scholarly endeavors over many years; and our wives, for not only gracefully surrendering family time so that the work might progress, but also hosting visits by the co-authors for periods of concentrated work.

# *Note on Citations of Classical Authors*

Where a standard division of a Greek or Latin author into books and chapters exists, we have used it, providing just the author's name when only one work is in question. For authors who have more than one established system for referencing, we have chosen the one that is in most common use. Thus passages of Strabo's *Geography* are cited by book number, chapter, and section number, e.g., Strabo 2.5.5, rather than by the page numbers of Casaubon or Almaloveen that appear in some editions. For the reader's convenience we have added page references to the translation in the Loeb Classical Library where one exists, and in the case of Ptolemy's *Almagest*, to Toomer's translation. Quotations are our own translations unless otherwise noted.

PTOLEMY'S

# *Geography*

# Introduction

Ptolemy's *Geography* is a treatise on cartography, the only book on that subject to have survived from classical antiquity. Like Ptolemy's writings on astronomy and optics, the *Geography* is a highly original work, and it had a profound influence on the subsequent development of geographical science. From the Middle Ages until well after the Renaissance, scholars found three things in Ptolemy that no other ancient writer supplied: a topography of Europe, Africa, and Asia that was more detailed and extensive than any other; a clear and succinct discussion of the roles of astronomy and other forms of data-gathering in geographical investigations; and a well thought out plan for the construction of maps.

Ptolemy himself would not have claimed that the *Geography* was original in all these aspects. He tells us that the places and their arrangement in his map were mostly taken over from an earlier cartographer, Marinos of Tyre. Again, Ptolemy comprehended fully the superior value of astronomical observations over reported itineraries for determining geographical locations, but in this he was, on his own admission, anticipated by other geographers, notably Hipparchus three centuries earlier.[1] Even so, he was too far ahead of his time in maintaining this principle to be able to follow it in practice, because he possessed reliable astronomical data for only a handful of places.

But in the technique of map-making Ptolemy claims to break new ground. He introduced the practice of writing down coordinates of latitude and longitude for every feature drawn on a world map, so that someone else possessing only the text of the *Geography* could reproduce Ptolemy's map at any time, in whole or in part, and at any scale. He was apparently also the first to devise sophisticated map projections with a view to giving the visual impression of the earth's curvature while at the same time preserving to a limited extent the relative distances between various localities.

At the very outset of the *Geography*, Ptolemy describes his subject as "an imitation through drawing of the entire known part of the world together with the things that are, broadly speaking, connected with it," and the work's Greek

[1]The geographer Strabo (1.1.12, Loeb 1:23–25) also ascribes to Hipparchus the opinion that the relative positions of widely separated places must be determined by astronomical observation.

title, *Geōgraphikē hyphēgēsis*, can be rendered as "Guide to Drawing a World Map." The core of the *Geography* consists of three parts necessary for Ptolemy's purpose: instructions for drawing a world map on a globe and on a plane surface using two new map projections (Book 1.22–24), a catalogue of localities to be marked on the map with their coordinates in latitude and longitude (2.1–7.4), and a caption or descriptive label (*hypographē*) to be inscribed on the map (7.5). As a supplement Ptolemy adds instructions for making a picture of a globe with a suitable caption (7.6–7), and describes a way of partitioning the known world into twenty-six regional maps, with a detailed caption for each (Book 8). The introductory chapters (1.1–21) set out fundamental principles for obtaining the data on which the world map is to be based, and necessary conditions for a good map projection; Ptolemy devotes much space here to criticism of his predecessor, Marinos.

For most modern readers, the parts of greatest interest will be those treating the theoretical questions and the relationship of Ptolemy's work to that of his predecessors. The enormous catalogue of localities and their coordinates is chiefly of concern to specialists in the geography of various parts of the ancient world, for whom an edited Greek text is indispensable. Accordingly, our translation omits the geographical catalogue and the captions for the regional maps, although we have provided a specimen of each.

The plan of the *Geography* is, for such a long work, very simple; yet certain of its features have turned out to be pitfalls. First, there is Ptolemy's characteristically parenthetic style of writing. His thoughts are continually being suspended partway through by qualifications and digressions, and completed only much later, which tends to give rise not only to long, elaborately nested sentences, but also to paragraphs of reasoning that sometimes extend over several chapter divisions.[2] The reader who is not prepared for Ptolemy's fondness for suspension and resumption of argument may be led to suspect that the text has been subjected to extensive interpolations, or even that Ptolemy did not know his own mind.[3]

Another serious difficulty is presented by the chapters in Books 7 and 8 that are entitled *hypographē*, a word that has usually been interpreted as "description" (of a map). If they are read in the same way as the other narrative parts of the *Geography*, that is, as Ptolemy speaking to the reader, then it is not easy to see the reason for their presence in the text. Historians have been taxed to explain why these chapters repeat material presented elsewhere in the *Ge-*

[2]It is not clear to what extent Ptolemy was himself responsible for the traditional division of the *Geography*'s text into chapters. Toomer (1984, 5) has raised doubts about whether the chapter divisions of the *Almagest* are Ptolemy's; and some of the chapter titles in the *Geography* break the text in awkward places or inadequately describe the contents.

[3]Polaschek (1959) is particularly given to hypotheses of this kind.

*ography*, and why the *hypographai* for the twenty-six regional maps express the locations of cities according to a system of coordinates different from the longitudes and latitudes of the catalogue in Books 2–7.[4] These apparent redundancies and inconsistencies, together with variations in the order and contents of the *Geography* as it is presented in the medieval manuscripts, have given rise to theories of the work's origin that deny its integrity—for example, hypothesizing that Ptolemy wrote Book 8 long before the rest of the *Geography*, or even that the various parts of the work were originally separate compositions by perhaps several authors, united only in the Middle Ages under Ptolemy's name.[5] The scribes who furnished some of the medieval manuscripts of the *Geography* with maps comprehended the function of the *hypographai*, however: they rightly used them as captions to be "written below" (*hypographein*) the maps. In these chapters Ptolemy is not addressing the cartographer; rather the cartographer is addressing the public. The *hypographai* should be understood as if within quotation marks, as part of the map-making kit.

A third obstacle is the degree to which different manuscripts diverge in the versions that they present of parts of the *Geography*, which is one reason why no satisfactory edition of the whole text of the work has been achieved in modern times. There can be no doubt that the *Geography* was badly served by its manuscript tradition; the most conscientious scribe was certain to introduce numerous errors in copying its interminable lists of numbers and place names, and some copyists did not resist the temptation to "emend" the text. The instability of the textual tradition chiefly affects the geographical catalogue and the captions for the regional maps.

Our translation attempts to redirect the reader's focus away from the topographical details of the map, as represented in the catalogue and the regional captions, to where we believe it belongs, which is on Ptolemy's exposition of the theory and method of cartography. We accepted as a working hypothesis that the *Geography* as it has come down to us is a coherent, intelligent, and logically organized treatise that forms an integral part of Ptolemy's scientific oeuvre and belongs to an identifiable stage in the development of his thought. The experience of interpreting and annotating the work has only confirmed our belief that this is the appropriate way to approach it.

## *What Ptolemy Expected His Reader to Know*

In addition to the circumstances that we have already described that have tended to obscure Ptolemy's purpose in writing the *Geography*, the book presents diffi-

[4]On the regional *hypographai* in Book 8 see, for example, Berger 1903, 643–644; and on 7.7 (the *hypographē* to the picture of the ringed globe), Neugebauer 1959, 29.

[5]For the first theory, see Schnabel 1930; for the second, Bagrow 1943.

culties for the modern reader that would not have been felt by readers of his own time. Sites to which he refers, which would have been instantly recognized by his contemporaries as thriving emporia and capitals of great kingdoms, are to most of us only names in a long list of places we have, at best, only read of as archaeological sites. Or, if they are known to the modern world, they often come to us cloaked in unrecognizable names, such as "Lake Maiōtis" for the Sea of Azov or "Taprobanē" for Sri Lanka. We have tried to lessen this last difficulty by providing the modern equivalents of places mentioned (when they can be identified) in the Geographical Index (Appendix H).

But Ptolemy writes against the background not only of a world that has vanished but also of a set of assumptions about the cosmos and its mathematical description, some of which are as foreign to the modern reader as are most of the localities he mentions. Accordingly, this section reviews the most important of Ptolemy's cosmographical presuppositions and their meanings, drawing where possible on Ptolemy's own treatment of these topics in his earlier astronomical treatise, the *Almagest*. We also discuss here the units of distance measurement and ways of describing directions that occur in the *Geography*.

*The Terrestrial and Celestial Spheres*

Ptolemy assumes that the reader understands and accepts the two-sphere model of the cosmos, that is, the geometrical conception of the heavens as an immense sphere that rotates daily around an axis through its center, with this center occupied by a second sphere, that of the earth (Fig. 1). The stars are thought of as fixed to the surface of the outer sphere, which is so vast that, as Ptolemy says in the *Almagest* (1.6, Toomer 43), "the earth has, so far as the senses can per-

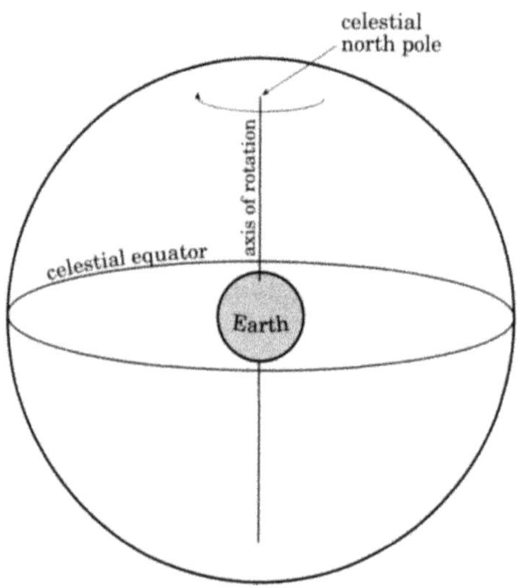


FIG. 1. The two-sphere model

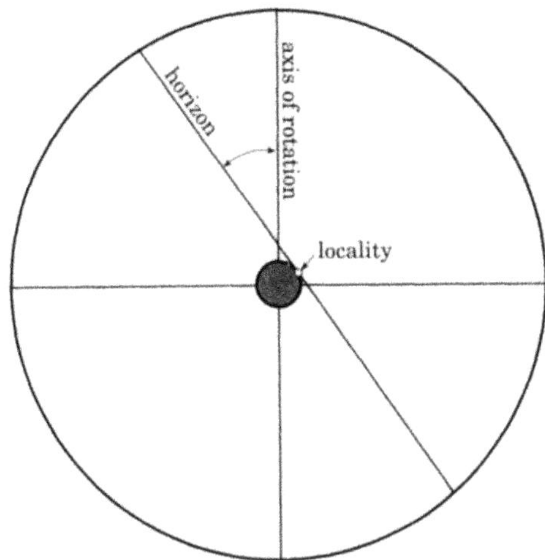


FIG. 2. Celestial sphere and local horizon of a terrestrial observer

ceive, the relation of a point to the distance to the sphere of the so-called fixed stars." The intersections of the axis of rotation with the sphere of the fixed stars define the north and south celestial poles, and, with respect to these directions, the daily rotation of the heavens is in a direction from east to west (i.e., clockwise if we imagine ourselves viewing the celestial sphere from outside and above its north pole). As a result of this daily rotation, the stars fixed to the surface of the celestial sphere trace out parallel circles, all centered on the poles, and the largest of these parallel circles is the *equator*, which is defined by the plane through the center of the cosmos and perpendicular to the axis.

*The Horizon*

Since the earth is a sphere, each locality on its surface admits a tangent plane, known as its horizon plane. However, Ptolemy reminds his readers in the *Almagest* (1.6, Toomer 43) that one of the reasons for regarding the earth as being so small relative to the cosmos is that the horizon plane seems to divide the celestial sphere into two exactly equal parts and could, therefore, be taken as passing through the center of the earth. The horizon, then, is another great circle of the cosmos, but it must not be thought of as rotating, for the earth did not rotate in the Ptolemaic cosmos. Rather, for a particular locality, the horizon is imagined as being fixed and therefore as making a fixed angle of inclination with the axis of rotation of the celestial sphere (Fig. 2).

*Parallels and Latitude*

This angle of inclination, known to Ptolemy as the *latitude* of a locality, varies with the location of the observer and determines which stars are capable of being seen. An observer at the north or south poles, whose latitude is 90°, would

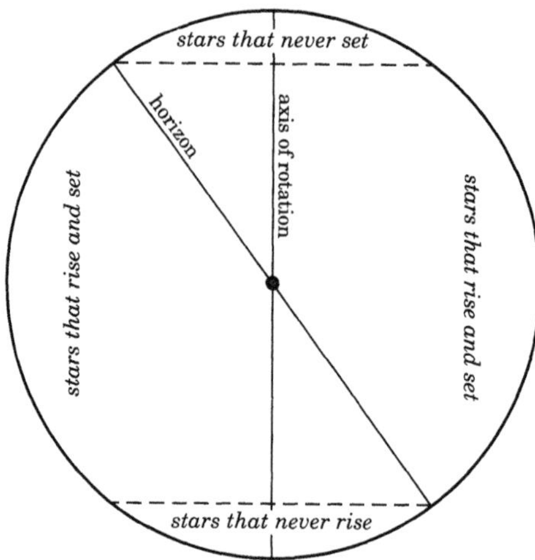


FIG. 3. Classes of stars on the celestial sphere

find that the equator coincides with the horizon and that stars north of the equator are always visible at night, and those south of the equator are always invisible. When the inclination is 0° (i.e., the horizon plane is parallel to the axis of the cosmos), the observer is on the earth's equator, both celestial poles are on the horizon, and all stars rise and set—each spending as much time above the horizon as below.

Ptolemy assumes, however, that his reader is at an intermediate latitude of the northern hemisphere, and for such a person the stars fall into three groups: stars that never set but are always above the horizon; stars that rise and set, and therefore are sometimes visible and at other times invisible, and stars that never rise and therefore are always invisible (Fig. 3). Separating these three groups of stars on the celestial sphere are two parallel circles of equal size. The one, to the viewer's north, separates the stars that never set from those that set and rise and is known as *the greatest of the always visible circles*. The other, to the viewer's south, separates the stars that never rise from those that set and rise and is known as *the greatest of the always invisible circles*. The two points where these circles touch the horizon mark due north and south for the observer, and the intersections of the equator with the horizon mark the points due east and west of the observer. Thus for the ancient geographers, geographical directions were in the first instance defined astronomically.

As one proceeds northward from the equator, the circle of ever-visible stars grows until, at the north pole, it coincides with the horizon. Simultaneously, the circle of always invisible stars also increases. Consequently, it can be demonstrated that locality *A* is north of locality *B* if some star in the northern hemisphere is always visible at *A* but rises and sets at *B*, or if some star that cannot be seen at *A* rises and sets at *B*.

These are just two astronomical criteria among many that may be used to judge how far north of the equator a locality is.[6] Because all these phenomena remain unaltered if one travels due east or west on the earth's surface, they define a *parallel* of latitude, that is, a circle on the terrestrial sphere parallel to the equator. The concept of identifying the phenomena characteristic of all localities having the same latitude, i.e., lying along the same parallel, had been known to geographical writers since the fourth century B.C.,[7] and Ptolemy specifically refers to it at several places in the *Geography* (1.2, 1.7, and 1.9). In *Almagest* 2.1 (Toomer 75–76) he lists as being among the more important phenomena characteristic for a latitude:

1. the elevation of the north or south celestial pole above the horizon;
2. whether, at any time during the year, the sun passes directly overhead;
3. the ratios of an upright stick (*gnōmōn*) to its shadow on the longest and shortest days of the year, as well as on the equinoxes; and
4. the amount by which the longest day of the year exceeds the equinoctial day, or equivalently, the ratio of the longest day of the year to the shortest, or simply the length of the longest day, measured in uniform time units.

In *Almagest* 2.6 (Toomer 82–90) Ptolemy adds two further phenomena to this list:

5. whether shadows in a given locality can point both north and south at different times of the year; and
6. which stars are always visible, which stars rise and set, and which stars can be directly overhead.

Phenomena (2) and (5) determine the latitude only within certain bounds. However, given any one of (1), (3), (4), and (6), we can determine the latitude and all the other phenomena, so that it is sufficient to specify any one of these three for a given locality. Ptolemy's basic datum is often the length of daylight; hence his principal parallels are chosen at constant increments of longest day. The latitudes corresponding to the regular sequence of increments in daylight are not equally spaced, but become more crowded the further we get from the equator. For this reason Ptolemy uses quarter-hour increments until he reaches the parallel for which the longest day is 15½ hours, and increments of half an hour thereafter until he reaches the parallel that he believes marks the northern limit of the known world, where the longest day is twenty hours. Some of

[6] Analogous rules apply to places south of the equator (none are invoked in the *Geography*).

[7] The traveler Pytheas of Massalia (c. 330 B.C.) reported polar elevations and lengths of longest day for several of the places in northwestern Europe that he claimed to have visited; see Dicks 1960, 180, 185–187.

Ptolemy's principal parallels, including those that mark the southern and northern limits of the part of the world covered by his map, are shown in Figure 4.

Ptolemy's highlighting of a sequence of unequally spaced parallels defined by the maximum length of day instead of parallels at uniform intervals of, say, 5° seems awkward from a modern perspective, but reflects the traditional practice of Greek geography. Earlier writers often made use of a division of the Greco-Roman world into latitudinal strips, or *klimata* (sing. *klima*), such that within each *klima* the maximum length of day was supposed not to vary significantly. (*Klima* means "inclination," signifying the angle between the axis of the celestial sphere and the plane of the horizon.) The lists of *klimata* that are found in various classical authors vary in the range of latitudes that they cover, although the number of *klimata* was by convention seven, counted from south to north. Ptolemy generally eschews the *klimata* in his own astronomical and geographical writings, but they figured in the work of his predecessor Marinos.

*Meridians and Longitude*

Intervals of time are also fundamental in the division from east to west. If we imagine a plane containing the north-south axis and passing through a locality on the earth's surface, this plane will intersect the terrestrial sphere in a great circle called a *meridian*. All places on the same meridian will observe the sun's

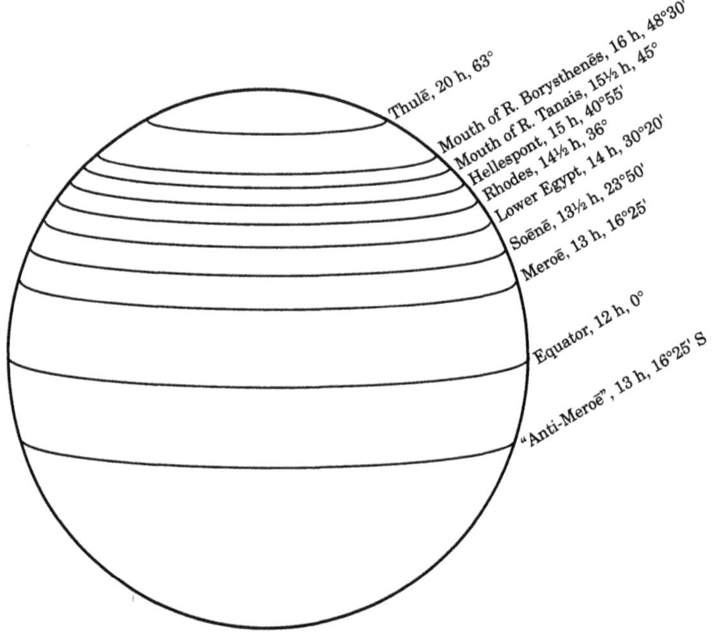


FIG. 4. The principal parallels defined by greatest length of daylight

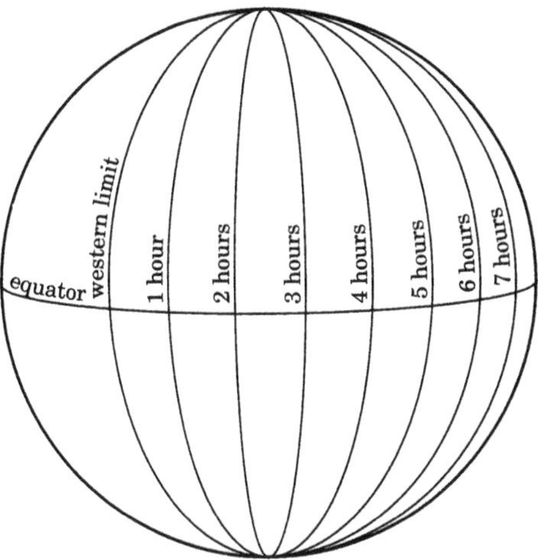


FIG. 5. The principal meridians defined by difference in local time

noon crossing of the meridian plane at the same time. Whereas latitude is readily defined by taking the arc of any meridian cut off between a special parallel, the equator, and a given parallel, there is no natural counterpart of the equator among the meridians from which the *longitude*, or angle to the other meridians, should be measured. By convention Ptolemy chooses to count longitudes eastward from the meridian at the western limit of the world known to him, and he writes (1.23) that "it is appropriate to draw the meridians at intervals of a third of an equinoctial hour," that is, at intervals of 5°." Thus it is fundamentally a net of time, not of degrees, that Ptolemy casts over the earth (Fig. 5).

*The Ecliptic*

An important great circle on the celestial sphere, and rotating with it, is the *ecliptic*, which Ptolemy refers to either as "the zodiacal circle" or as "the circle through the middle of the signs [of the zodiac]." The sun traverses this circle annually at an average rate of just under a degree each day, from west to east relative to the stars—that is, opposite to the daily rotation of the celestial sphere.

Since the ecliptic is the central circle of the belt of signs making up the zodiac, it inherits that belt's division into signs—the familiar Aries, Taurus, Gemini, etc., shown in Figure 6. The annual eastward progress of the sun is counterclockwise in the diagram. The signs are each 30° in length, and so coincide only approximately with the constellations for which they are named.

Since the ecliptic is a great circle like the equator, it intersects the equator in two diametrically opposite points: the beginning of Aries, where the sun is at the spring equinox, and the beginning of Libra, where the sun is at the autumnal equinox. The ecliptic is tilted at an angle of about 24° with respect to the celestial equator, and so there is a most northerly point on the ecliptic, located at the beginning of the zodiacal sign of Cancer, and a most southerly point, at the beginning of the sign of Capricorn. The circles on the celestial sphere that are parallel to the equator and that pass through these two points are known, respectively, as the Tropic of Cancer (or Summer Tropic) and the Tropic of Capricorn (or Winter Tropic). As the sun travels annually around the ecliptic, it moves alternately north and south of the equator, with the two tropic circles as the limits of this motion (Fig. 7).

The center of the earth is the center of the cosmos; hence it may be used to define "down" in the cosmos as toward the center of the earth and "up" as away from the center. With this understood, one can imagine for the equator and tropic circles on the celestial sphere a corresponding circle directly below it on the earth, and we shall follow the Greeks in using the same names for the terrestrial circles as for their celestial counterparts. The terrestrial tropics are limiting circles for one of the varieties of astronomical phenomena used to determine latitude: only for observers in the belt between them does the sun pass directly overhead in the course of the year.

Another pair of circles closer to the terrestrial poles have a corresponding limiting role for a different latitudinal phenomenon. For observers at these circles, the length of the longest day of the year just reaches its greatest possible value, twenty-four hours, so that between these circles and the poles there will be some days of the year when the sun never sets. The limiting circle surrounding the north pole is the *arctic* circle, and its southern counterpart is the *antarctic* circle. Each is as far from its pole as the tropics are from the equator.

### *Climatic Zones*

Although the various circles on the celestial sphere are primarily of astronomical significance, some ancient geographers used the corresponding circles on

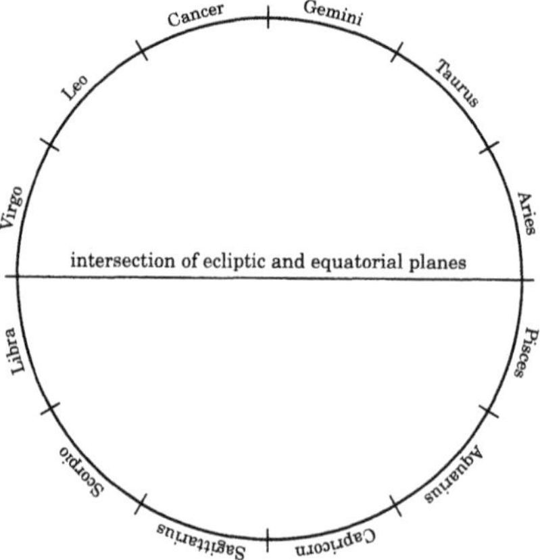


Fig. 6. The ecliptic and zodiacal signs

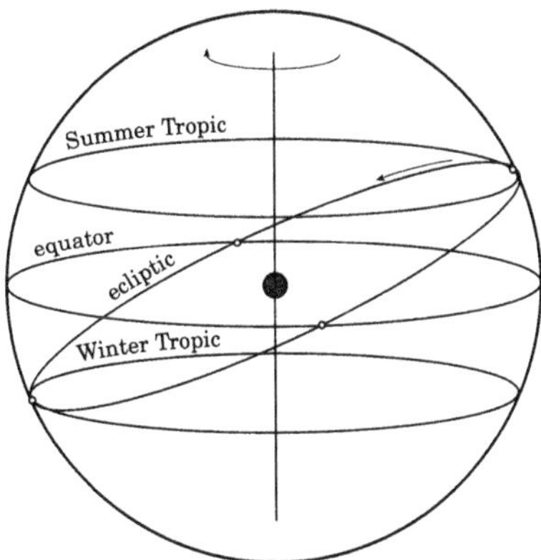


FIG. 7. The ecliptic and tropic circles

the earth's surface to divide the earth into zones with geographical, and even climatic, significance. Thus, according to Aristotle (*Meteorology* 2.5, 362a32, Loeb 179–181), there were two "frigid" zones (one north of the arctic circle and one south of the antarctic circle), two "temperate" zones (between the frigid zones and the two tropics), and a torrid zone (located between the tropics). It appears from 1.7 that Marinos set the limits of the torrid zone at a bit more than 12° north and south of the equator—as did Posidonius before him.[8] Ptolemy occasionally makes use of the principle that climate (including the range of plant and animal life and the appearance of the human inhabitants) is dependent on latitude to deduce that localities sharing the same climate must be at approximately the same distance from the equator.

Ptolemy also uses an even simpler division of the earth's surface, based on shadows rather than *klimata*. At localities between the two tropics the noon sun would be, according to the time of the year, sometimes to the north and sometimes to the south of the zenith, so that the corresponding shadows of a vertical rod (*gnōmōn*) would, during the course of the year, point north on one day and south on another. (Thus the regions are referred to as *amphiskian*, for the Greek word signifying that the shadows point in both directions, north and south, during the course of a year.) For persons exactly on the tropic circles, the noon shadows would point always north or always south, with the exception of one day of the year on which there is no noon shadow. At localities between the tropics and the arctic or antarctic circles, noon shadows will always point north or always point south. Such localities are known as *heteroskian*. Finally, at localities between the poles and the arctic or antarctic circles, there will be a part of the year during which the *gnōmōn*'s shadow makes a complete circuit around it. These localities, called *periskian*, play no role in the *Geography*.[9]

[8]Strabo (2.2.1–2.3.1, Loeb 1:361–371) has a very interesting discussion of geographical zones.

[9]See the discussion in *Almagest* 2.6 (Toomer 89–90) and Strabo 2.5.43 (Loeb 1:517–521).

*Degrees*

Ptolemy makes use of the *degree* as a unit for measuring arcs along meridian circles and parallels of latitude. This unit, which had its origin in the Babylonian practice of dividing both the day and the zodiac into 360 equal parts, was already being applied by the Greeks to circles on the celestial and terrestrial spheres in Hipparchus' time. Ptolemy, however, seems to have been the first geographer to establish a uniform coordinate system in degrees for specifying precise positions on the earth's surface. This system was devised on analogy with a convention astronomers had long been using to specify positions of stars and planets on the celestial sphere by two numbers: a *latitude* ("breadth") above or below the ecliptic, and a *longitude* ("length") measured along the ecliptic from a conventional zero point. For geographical purposes the equator replaces the ecliptic, and Ptolemy measures latitude north or south from the equator to a locality along a meridian circle, and longitude along the equator between that meridian and the meridian passing through the westernmost place on his map (the Islands of the Blest). Compared to the divisions of the globe based on celestial phenomena, the coordinates of latitude and longitude had the practical advantage for the cartographer of precision and uniformity of units. Nevertheless Ptolemy preferred that the finished map and its captions should express everything in terms of hour divisions and the other fundamental, astronomically defined circles.

*Units of Distance*

Measured linear distances from place to place were expressed in several different kinds of unit in the various sources on which Marinos and Ptolemy drew. The most important of these units was the *stade*, the standard unit of terrestrial distance in classical geography, which was probably understood by Ptolemy and his predecessors as a distance amounting to approximately 185 meters.[10] Stades could be converted into degrees according to the assumed equivalence of 500 stades to one degree measured along the equator or along a meridian. Distances from Roman sources, for example those pertaining to the Roman roads, would be expressed in the Roman mile (approximately 1.48 kilometers), which was usually treated as interchangeable with eight stades. In Egypt distances could be stated in the *schoinos*, which Ptolemy takes to be thirty stades. For the roads of the Parthian Empire the old Persian *parasang* was used; this was near enough in length to the *schoinos* so that in Ptolemy's sources the Egyptian name is substituted, and the same ratio is applied to convert to stades.

[10]There has been much disagreement concerning whether there was a single standard stade employed by all the geographical writers, and how large it was. We agree with Dicks (1960, 42–46), Engels (1985), and Pothecary (1995, 50–51) that they all used—or at least believed that they were using—the so-called Attic stade.

*Directions*

Ptolemy alludes to two ways of describing directions of travel, one based on the points of the horizon where the sun rises and sets, the other based on conventional names of the winds that blow from various directions. On the vernal and autumnal equinox, the sun is seen to rise due east of an observer, and to set due west. Hence these directions are sometimes called the directions of equinoctial sunrise and sunset. During the half of the year when the sun is north of the equator, which includes the summer for the northern hemisphere, the points of sunrise and sunset on the horizon are north of due east and west, reaching an extreme limit on the summer solstice; and similarly the rising and setting points are furthest south of due east and west on the winter solstice. Ptolemy refers to these directions as the directions of the sun's summer or winter rising or setting. In fact, they are not the same for observers at different latitudes: at the equator they are approximately 24° from due east and west, but the angles become larger as one moves further away either north or south. Ptolemy treats them, however, as being 30° from the east-west line regardless of the latitude; this is approximately correct for the latitude of Rhodes, which was traditionally thought of as the central east-west axis of the known world.

Additionally, Ptolemy and his sources use a scheme of twelve winds to specify directions. Four of these are equivalent to the cardinal directions, north, south, east, and west. The remainder are treated as equally spaced at 30° intervals between the cardinal directions, so that for Ptolemy the system based on the sun's rising and setting points is largely interchangeable with the system based on winds. The whole scheme is illustrated in Figure 8 (wind names in parentheses do not occur in the *Geography*). Note that the arrowheads indicate the direction of travel *toward* the designated wind, which is of course opposite to the direction from which the wind is supposed to blow.

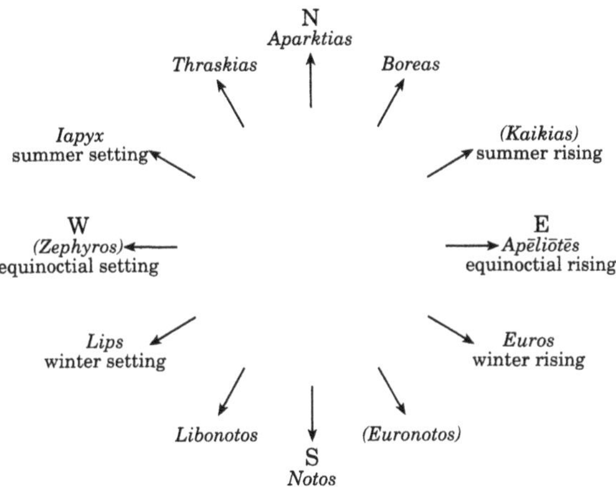


FIG. 8. Indications of direction by winds, sunrise, and sunset

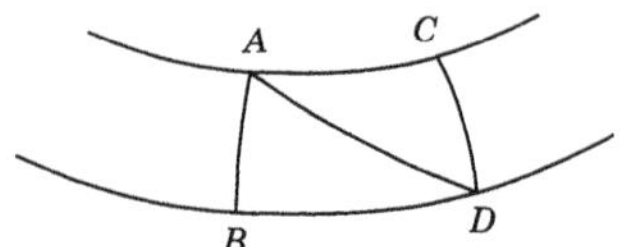


FIG. 9. Conversion of terrestrial distances to longitude and latitude

*Conversion of Distance Measurements to Degrees*

Ptolemy often has to translate a given interval between two localities, expressed as a number of units of distance in a particular direction, into the number of degrees of longitude between the meridians through the two localities and the number of degrees of latitude between their parallels. His procedure sometimes involves several stages.

a. If a locality $A$ is $s$ stades due north of another locality $B$, or vice versa, they lie along the same meridian (Fig. 9). Since a meridian is a great circle, Ptolemy uses the assumed equivalence of one degree with 500 stades along a terrestrial great circle. The difference in degrees between their latitudes is $s/500$.
b. If locality $A$ is $s$ stades due west of locality $C$, or vice versa, they lie along the same parallel (Fig. 9). Since a parallel is not a great circle (unless it happens to be the equator), Ptolemy has to find the number of stades corresponding to 1° along the parallel, which is in the same ratio to 500 as the circumference of the parallel is to the circumference of the equator. Suppose that the latitude of $A$ and $C$ is $\phi$ degrees. The circumference of the parallel at latitude $\phi$ is $\cos\phi$ times the circumference of the equator.[11] Hence the difference in degrees between the longitudes of $A$ and $C$ is $s/(500\cos\phi)$.
c. If $A$ is $s$ stades from $D$ in some intermediate direction (Fig. 9), we must analyze the interval between them into north-south and east-west components, $AC$ and $CD$ respectively. In doing this, Ptolemy neglects for the moment the sphericity of the earth; that is, he regards the spherical triangle $ACD$ as so small relative to the earth that it may be treated as a plane triangle with a right angle at $C$. Let the horizon angle between $AD$ and the parallel through $A$ (i.e., angle $CAD$) be $\theta$. Then the east-west component of the interval in stades is $s(\cos\theta)$, and the north-south component is $s(\sin\theta)$. Each component is separately converted to intervals in degrees as described above.

[11]Ptolemy would not have employed the modern trigonometrical functions *sin* and *cos*, but rather the "chord" function, which is the length of a chord subtended by a given angle in a circle of radius 60. A table of chords as a function of angles is presented in *Almagest* 1.11. What we call cos $\theta$, Ptolemy would have calculated as chord $(180° - 2\theta)/120$.

d. We have assumed that we know the direct-line distance between $A$ and $B$. Ptolemy recognizes, however, that distances estimated by travelers generally are longer than the most direct route. One reason for this was that the route taken was not always straight; for example, mariners would follow the outlines of bays rather than sail straight across. Moreover, distances expressed in stades were sometimes calculated from the time taken in making the journey by multiplying by an assumed ideal rate of travel, but this would lead to an exaggerated figure if there had been delays on the route. Ptolemy allows for these tendencies in a very arbitrary way, typically by reducing a reported stade distance by one-third. Thus Ptolemy's analysis of a reported interval from one place to another can often involve steps (d), (c), and (a) and (b), in that order.

## *The Place of the* Geography *in Ptolemy's Work*

Ptolemy (or, to give his full name, Klaudios Ptolemaios) was born about A.D. 100 and began his scientific career in the mid-120s, working in or near Alexandria in Egypt. He probably lived into the last quarter of the century.[12] Ptolemy's incitement to determine numerical coordinates for geographical locations throughout the known world may have come from the astronomical researches with which his scientific career began, and for which he is now best known.

We can see this origin in the *Almagest*, Ptolemy's great treatise on the mathematical theory of the motions of the heavenly bodies, which is generally regarded as his earliest major writing.[13] The *Almagest* is concerned with the apparent motions of the sun, moon, planets, and fixed stars, how to account for them quantitatively by means of models involving combinations of circular motions, and how to compute the instantaneous positions of the heavenly bodies and other celestial phenomena using tables based on these models. Geographical considerations arise in various ways in the execution of Ptolemy's project, most obviously in the fundamental problem of converting the recorded times of astronomical observations made in different places to Alexandria mean time. The same astronomical event will be observed in two places of different longitude at different intervals of time since the preceding local noons, and this difference is proportional to the difference in longitude between the two places. Moreover, ancient observers did not measure the times of observations in constant *equinoctial* hours after noon or midnight. Instead they divided the two intervals between sunrise and sunset and between sunset and sunrise into twelve equal *seasonal* hours, and described observations as having occurred at such-

[12]For a survey of Ptolemy's life and works, see Toomer 1975.

[13]The *Almagest* was finished later than A.D. 147 (Hamilton et al. 1987); in it Ptolemy cites astronomical observations that he made from A.D. 127 on.

and-such an hour of day or of night. To convert a reported time in seasonal hours to the number of equinoctial hours since local noon, one had to know the length of the seasonal hour in equinoctial hours, which is a function of both the sun's position on the ecliptic (i.e., the time of year) and the observer's latitude.

If Ptolemy had worked only on the basis of his own observations, he would still have needed to know the latitude of his locality, Alexandria. Since, however, he also used older observations made in a few other places, he needed values not only for the latitudes of these sites, but also for the differences between their longitudes and the longitude of Alexandria. Similarly, anyone else not living at Alexandria who wished to use Ptolemy's tables would have had to know his own latitude and relative longitude in order to convert his local time in seasonal hours to the uniform time of the tables, and vice versa.

In the *Almagest*, Ptolemy's treatment of matters related to the observer's geographical position is almost wholly theoretical. The relationship between longitude and time is a simple proportionality, and requires no special discussion.[14] For the more complex problems connected with latitude, Ptolemy designates a series of special parallels on the earth, computing for each parallel relevant astronomical data, including the tables of *oblique ascensions*, which are the basis for converting seasonal to equinoctial hours.[15] The complete list of parallels starts with the equator, and proceeds north at intervals such that the duration of daylight at the summer solstice increases by quarter-hours from 12 equinoctial hours at the equator to 18 equinoctial hours at 58° N, and then by larger time-intervals (because the parallels get closer together) to 24 equinoctial hours at the arctic circle (66°8'40" N). For the purposes of his tables, Ptolemy cuts this list down first to eleven parallels at intervals of half-hours of increase in longest daylight from the equator to 54°1' N (17 equinoctial hours), and later to just seven parallels (cf. Fig. 4) at half-hour intervals from 16°27' N (13 hours) to 48°32' N (16 hours).[16]

For most of these parallels, Ptolemy indicates a geographical location through which the parallel passes. In some instances this location is a city; for example, the parallel for 13½ hours is through the city Soēnē (modern Aswān). Other parallels are said to pass through less precise geographical features, or even

[14]Ptolemy states the rule (fifteen degrees of longitude correspond to one equinoctial hour) briefly in *Almagest* 2.13 (Toomer 130) and again at 6.4 (Toomer 282).

[15]*Almagest* 2.6–13. The oblique ascension associated with a given point on the ecliptic is the arc of the celestial equator that rises at the horizon of a given locality simultaneously with the arc of the ecliptic between the vernal equinoctial point, Aries 0°, and the given point. The interval in equinoctial hours between sunrise and sunset is proportional to the arc of the equator that rises above the horizon during that time, i.e., the difference between the oblique ascension of the point of the ecliptic diametrically opposite to the sun and the oblique ascension of the sun's position.

[16]In giving special prominence to these seven parallels, Ptolemy was following in an established tradition; see p. 10.

broadly defined districts; for example, that for 14 hours passes through Lower Egypt (the Nile delta). Apart from these latitudes, the only explicitly stated geographical data in the *Almagest* occur in the context of analyzing specific observations. Thus, Ptolemy gives Alexandria's latitude (30°58'), as well as latitudes and time differences from Alexandria for Babylon, Rhodes, and Rome.

The scarcity of geographical data in the *Almagest* is deliberate. At the end of the section in which he computes and tabulates the astronomical phenomena for his series of parallels, Ptolemy writes (*Almagest* 2.13, Toomer 122–130):

> What is still missing in the preliminaries is to determine the positions of the noteworthy cities in each province in longitude and latitude for the sake of computing the phenomena in those cities. But since the setting out of this information is pertinent to a separate, cartographical project, we will present it by itself following the researches of those who have most fully worked out this subject, recording the number of degrees that each city is distant from the equator along the meridian described through it, and how many degrees this meridian is east or west of the meridian described through Alexandria along the equator, because it was for that meridian that we established the times corresponding to the positions [of the heavenly bodies]. For the present, however, we take the [geographical] locations for granted.

The project of compiling a catalogue of important cities and their coordinates, which Ptolemy had not finished (and perhaps had not even begun) when he wrote this, was the germ from which the *Geography* grew. On the way, however, Ptolemy's scope broadened from the establishment of coordinates for a few hundred cities to a far more comprehensive codification of thousands of elements (towns, borders, natural features) of the entire known world; and his primary purpose shifted from compiling a table ancillary to his astronomical tables to laying down new and better foundations for drawing maps of the world.

Ptolemy did not, however, lose sight of his earlier plan. Among the roughly 8,000 localities in the huge catalogue of *Geography* Books 2–7, several hundred cities and towns were marked as being of particular importance;[17] and in the captions of the twenty-six regional maps (Book 8) Ptolemy listed these "Important Cities" with their positions translated into time units: the time difference from the meridian of Alexandria in equinoctial hours, and the length in equinoctial hours of the longest daylight. And when Ptolemy published a revision of the tables of the *Almagest* as a separate work, entitled the *Handy Tables*, he included in it a "Table of Important Cities," which presents substantially the same cities that he picked out in the *Geography*, with their longitudes and lati-

[17]The important cities originally seem to have been indicated by a special symbol in the margin, a notation that survives vestigially in at least one manuscript.

tudes in degrees extracted from the main catalogue of Books 2–7 and listed more or less in the order of Book 8.[18] Aside from this table, which is more an abridgment than a revision, the *Geography* appears to represent Ptolemy's final word on geographical questions.

## *Ptolemy's Evolving Conception of the World*

When Ptolemy wrote the *Almagest*, he accepted a geographical picture of the known, inhabited world (the so-called *oikoumenē*) that was not radically changed from that of Eratosthenes (third century B.C.) and Hipparchus (c. 140 B.C.). He accepted as a matter of course that the earth was spherical; *Almagest* 1.4 presents arguments on this point, but by Ptolemy's time scarcely any educated person would have seriously questioned it.

There is some reason to believe that at this stage Ptolemy accepted the estimate going back to Eratosthenes that the earth's circumference is approximately 250,000 stades, which was usually expressed by the equation of one degree of the earth's equator with 700 stades.[19] If, as we believe, one stade was approximately 185 meters, then Eratosthenes' measurement (which was based on heavily rounded data) was about 15 percent too large.

The whole of the *oikoumenē* fits inside one quarter of this sphere, bounded on the south by the equator and on the east and west by a single meridian circle (*Almagest* 2.1, Toomer 75). Ptolemy was willing to believe (*Almagest* 2.6, Toomer 83) that the regions along the equator had a habitable climate, less torrid perhaps than districts closer to the Tropic of Cancer because the sun was close to the zenith for a briefer part of the year; but it was his opinion that no one from the Greco-Roman world had ever been as far south as the equator, and that one could not trust tales purporting to describe what was found there. The southernmost locality to which Ptolemy refers is the island Taprobanē (Sri Lanka), which he situates on the parallel 4¼° north of the equator. No place is mentioned on the east coast of Africa further south than the Bay of Avalitēs (north of the Horn of Africa), and no place on the Nile further south than Meroē (between the junctions of the Blue Nile and the Atbara with the White Nile). At the

[18]The order in which the cities are listed in all three contexts (*Handy Tables*, *Geography* 2–7, and *Geography* 8) is determined first by Ptolemy's division of the world into the twenty-six maps, and subordinately by the logical order in which the features of each province are supposed to be drawn on the map. This fact establishes that the *Geography* must have taken its present form (if it had not actually attained its final state) before the *Handy Tables* were published.

[19]The evidence is that Ptolemy assumes smaller time differences between the meridians of Rome, Alexandria, and Babylon in the *Almagest* than in the *Geography*, roughly in the proportion that would result if the same stade distance had been converted to degrees of longitude using respectively 700 stades and 500 stades to the degree (Schnabel 1930, 219). On Eratosthenes' measurement of the size of the earth, see, e.g., Dicks 1971, 390–391.

northern extremity of the *oikoumenē*, Ptolemy states that the parallel 64½° north of the equator passes through "lands of the unknown Skythians," presumably in the Baltic regions. Parallels from 63° southward to 55° are associated in turn with the island of Thulē, the Hebrides, Ireland, and places in northern and central England. The inclusion of the British Isles and the mouth of the Rhine in Ptolemy's list of parallels is the only prominent reflection in the *Almagest* of geographical knowledge acquired since the beginning of the Roman Empire in the late first century B.C.

Between the *Almagest* and the *Geography*, Ptolemy wrote an important astrological treatise known as the *Tetrabiblos*, in which there is a chapter (2.3, Loeb 129–161) setting out his version of the traditional topic of astrological geography, correlating the supposed characteristics of various peoples with the influences of the zodiacal signs and the planets. Again Ptolemy situates the *oikoumenē* inside a half of the northern hemisphere, and he further partitions this into four quarters divided by a parallel passing through the Mediterranean and along a range of mountains extending eastward through Asia, and by a meridian passing through the Sea of Azov, the Black Sea, the Aegean, and the Red Sea. If this meridian was intended to bisect the *oikoumenē* longitudinally, then it may be inferred that the world known to Ptolemy did not yet extend eastward much beyond the Ganges, although the countries listed include Sērikē, the "Silk country" that represents the Chinese terminus of the Silk Road. In the southerly direction, Ptolemy now knows of Azania, a stretch of the East African coast south of the Horn that he was to situate just south of the equator in the *Geography*. Unfortunately, the seventy-two countries named in the *Tetrabiblos* are arranged in schematic groupings that correspond to their geographical locations only in a loose way, so that we cannot reconstruct an underlying "map."

The *oikoumenē* portrayed in the *Geography* is more extensive than it is presented, not merely in Ptolemy's earlier writings, but in any other classical text before or after Ptolemy, except for those few authors who adapted Ptolemy's work. By this stage Ptolemy was convinced by investigations that are otherwise unknown to us (and of which he gives no details) that the earth was a smaller globe than Eratosthenes had thought, so that only 500 stades corresponded to one degree of the equator, and the earth's circumference amounted to 180,000 stades.[20] Hence in contrast to Eratosthenes' estimate, Ptolemy's is about 18 percent too small.[21] His *oikoumenē* still fitted within the 180° of longitude

[20]Ptolemy also uses this smaller value for the size of the earth without comment in the *Planetary Hypotheses*, an astronomical work written after the *Handy Tables*; see Goldstein 1967, 11.

[21]It is often stated in modern discussions that Ptolemy took his figure of 180,000 stades for the circumference from a lost geographical work of Posidonius (first century B.C.). Ptolemy does not say so, and the only ancient source that appears to ascribe the number to Posidonius (Strabo 2.2.2, Loeb 1:361–365) contains a serious numerical inconsistency at just this point (Taisbak 1974). The

bounded by a single meridian circle, but only just; and it now stretched from the old northern limit at 63° to a southern limit more than 16° south of the equator. One of the more remarkable features of the map he draws inside this frame is that most of the edge consists of land, not ocean: Ptolemy was one of the few ancient geographers willing to admit that the theoretically habitable land mass of the world extended indefinitely beyond the limits of knowledge of his time.[22]

Ptolemy's *oikoumenē* is divided into three great continents, Europe, Libyē (our Africa), and Asia. To an eye accustomed to modern maps of the world, Ptolemy's Europe is the most instantly recognizable continent. The outline of the European mainland is complete as far north as the east coast of the Baltic. Distortions of direction and scale are obvious in the more remote parts toward the north and west, as in the outlines and relative positions of the British Isles; and even in the Mediterranean there is a surprising error of orientation in the shape of Italy. The accuracy of the Mediterranean and Red Sea coasts of Ptolemy's Libyē falls off somewhat as the Horn of Africa is rounded, but it is the Atlantic coast, with its straight north-south orientation terminated by a bend toward unknown lands to the southwest, that renders this continent strangely unfamiliar. Asia exhibits greater and greater distortions as one progresses further east, the most obvious faults being the north-south compression of the Indian subcontinent so that its western coast is made to run parallel to the equator, and the exaggerated size of the island of Taprobanē (Sri Lanka). At the eastern edge, where the lands represent central China and Southeast Asia, it is virtually impossible to identify any of the features on Ptolemy's map with real counterparts. At his eastern limit Ptolemy draws the coast of Asia as turning south and then west, eventually to join the east coast of Africa, thereby making the Indian Ocean a vast enclosed sea unconnected with the Atlantic Ocean.[23]

---

assertion (no less common) that Eratosthenes' measurement was remarkably accurate and Ptolemy's grossly in error results from supposing that the geographers' stade was much smaller than the Attic stade.

[22]Compare Strabo, writing a century and a half before Ptolemy, who maintains (1.1.8 and 2.5.5, Loeb 1:17–19 and 431–433) that the known *oikoumenē* is entirely, or almost entirely, surrounded by sea; and similarly, in the mid-first century A.D., Pliny the Elder (2.166–170, Loeb 1:301–305).

[23]Claims that mariners from Egypt or Spain had succeeded in circumnavigating the southern part of Africa were typically met with disbelief in antiquity (Herodotus 4.42, Loeb 2:239–241 and Strabo 2.3.4, Loeb 1:377–385). Hipparchus (cf. Strabo 1.1.8–9, Loeb 1:17–19) and Polybius (3.38, Loeb 2:89) had previously considered it possible that the Atlantic and Indian Oceans did not join south of Africa. Ptolemy does not actually provide coordinates for the coast of the unknown land linking Africa and Asia, but he refers to it verbally in 7.3, 7.5, 7.7, and 8.1.

## *Marinos and Other Sources*

For all his disagreement with his predecessor concerning points of method and detail, Ptolemy ungrudgingly acknowledges that the collection of geographical data presented in the *Geography* is substantially the work of Marinos of Tyre. As he tells us in 1.4–6, the cartographer's task is not to gather and digest afresh all the information that is to go into his map, but to take as his starting point the most recent comprehensive and competent work of the same kind, correcting and augmenting it using his critical skills and the most up-to-date specialized sources; and in Ptolemy's time it was Marinos' map and geographical writings that best represented the current state of knowledge.

It is from the *Geography* alone that we know of Marinos' existence and can reconstruct some aspects of his work.[24] Most of what Ptolemy has to say about him is by way of exposing his faults; but then Ptolemy expected that his reader would be able to consult Marinos' works and judge them for himself. Ptolemy's treatment of Marinos is not altogether unlike his treatment of Hipparchus in the *Almagest*: the mistakes of both his predecessors seemed deserving of careful exposition precisely because of the stature of their overall achievement. After all his criticisms, Ptolemy professes that his intention is "to preserve [Marinos'] opinions [as expressed] through the whole of his compilation, except for those things that need some correction" (1.19).

Although Marinos is first introduced to the reader as "the latest [author] in our time to have undertaken this subject" (1.6), the manner of Ptolemy's references to him strongly suggest that he was dead when Ptolemy undertook the *Geography*, and that some time had elapsed since his "final publication" (1.17): enough, at least, so that Ptolemy could write of discrepancies between Marinos' work and "the reports of our time." Ptolemy includes in his map features, presumably taken over from Marinos, that reflect the state of the Roman Empire about the first decade of the second century A.D., whereas there are extremely few features that came into existence after about A.D. 110.[25] Moreover, the latest explorations of the interior and east coast of Africa on which Marinos based his

[24]The tenth-century Arabic historian al-Mas'ūdī claimed to have seen a "book of *Geōgraphia* of Marinos," which contained maps (*Kitāb al-Tanbīh wa'l-ishrāf*, ed. de Goeje, p. 33; trans. Carra de Vaux, p. 53), but this is likely to have been a reconstruction from Ptolemy's text rather than an original work of Marinos. Elsewhere in the same book al-Mas'ūdī asserts that Marinos lived in the reign of the emperor Nero, i.e., A.D. 54–68 (ed. de Goeje, p. 127, trans. Carra de Vaux, p. 178). It is hard to imagine the source for this date, which is probably about half a century too early. Wieber (1995) surveys these and other Arabic references to Marinos, concluding that all Arabic knowledge of Marinos' works derived from Ptolemy.

[25]Honigmann (1930, 1767–1768) pointed out the presence in the *Geography* of place names reflecting Trajan's Dacian campaigns (which ended in A.D. 107), but none from the Parthian campaigns that began in 114. Desanges (1964, 40–41) established a similar terminal date of 110 for Ptolemy's description of north Africa.

estimates of the southern extension of that continent appear to have taken place in the second half of the first century.[26] We will therefore not be far off the mark if we situate Marinos' activity in the years about A.D. 100.

That Marinos produced an actual map of the *oikoumenē* seems clear from Ptolemy's criticism of his choice of projection in 1.20, and besides, it is difficult to imagine how Ptolemy could have constructed his own map without having access to the map of Marinos. For the most part, however, Ptolemy directs his attention to the series of writings that Marinos published on various aspects of the map. Ptolemy's references tend to be vaguer than we might wish, because he presumed that Marinos' writings would be accessible to his readers. In 1.6 he writes of Marinos' many "publications (*ekdoseis*) of the revision of the geographical map," which might mean either a number of reeditions of a major cartographical treatise or a series of bulletins setting out corrections to an initial version of the treatise or of the map itself. The next sentence seems to support the first of these interpretations, since it mentions Marinos' final "composition" (*syntaxis*) as a possible, though unsatisfactory, basis for drawing the map of the world.[27] Again in 1.17 Ptolemy attributes some of Marinos' inconsistencies to the abundance of information in his "compositions" and their being "split up" (*kechorismenon*). By this last word Ptolemy means that Marinos' "composition" was divided into sections devoted to single kinds of geographical data or relationships rather than proceeding region by region through the *oikoumenē*. We can identify some of these sections from the survey of Marinos' inconsistent statements in 1.15–16.

One part dealt with the identification of localities that were "oppositely situated" (*antikeimena*), a technical term that Ptolemy defines in 1.4 as being "on a single meridian." He adds that these places had been so identified by the observation that the sail from one to another was effected by the north or south wind; and in fact, all the instances cited from Marinos are of localities on the coasts of the Mediterranean and its islands.

In another part, which Ptolemy calls "the division of the *klimata* and of the hour-intervals," Marinos located places within latitudinal belts called *klimata* or within longitudinal sectors called *hour-intervals*. An hour-interval was the part of the globe bounded by two meridians separated by 15° of longitude, so that local noon (when the sun crosses the observer's meridian) would take place

[26]Desanges (1978, 197–213) gives plausible arguments for dating the expedition of Julius Maternus to Agisymba to about A.D. 90, and that of Septimius Flaccus a few years earlier. The voyage of Dioskoros, which extended knowledge of the African coast beyond Rhapta to Cape Prason, was known to Marinos but not to the author of the *Periplus of the Erythraean Sea*, who wrote about the middle of the first century (see p. 27 n. 32 below). The *Periplus* is also much vaguer than Marinos concerning the south coast of Asia beyond the Ganges.

[27]It is worth recalling that the original title of Ptolemy's *Almagest*, a comprehensive treatise in thirteen books, is "Mathematical *Syntaxis*."

one equinoctial hour earlier at the eastern edge of the hour-interval than at its western edge. The "description of the parallels" (1.15) seems to have been a treatment of circles of latitude separate from the section on the *klimata*. Presumably this part contained lists of localities that were supposed to lie exactly on, or slightly to the north or south of, certain significant parallels.

A section that Ptolemy calls "the definition of the boundaries" (1.16) apparently took up in turn each of the major regions and provinces, and described its outline in relation to the other regions and bodies of water that neighbored it to the north, east, south, and west. This is a structural device that Ptolemy imitated in the catalogue of the *Geography*, but it does not seem that Marinos provided the precise quantitative descriptions of each coast and boundary comparable to Ptolemy's lists of coordinates. Even so, the "provinces and satrapies" into which Ptolemy divided the known world for the purposes of his geographical catalogue are likely to correspond in large part to Marinos' regions.

Ptolemy's chapters on the latitudinal and longitudinal extent of the *oikoumenē* give us fascinating glimpses of the varied informants on whom he and Marinos drew for the more remote parts of the world: merchants, mariners, and soldiers. We know far less about the sources for the more accessible regions, such as the provinces of the Roman Empire itself, precisely because Ptolemy is willing to take over these parts from Marinos on trust. In this respect Ptolemy is following the practice of earlier geographers in thinking of the world map as a traditional rather than a personal production. One had to justify any innovations one was imposing on the inherited picture; but there was no need to cite evidence for whatever was left unaltered. Ptolemy's allusion to Marinos' "many publications of the revision of the geographical map" suggests that Marinos, too, thought of himself as a corrector rather than a creator.

Hence, although much effort has been expended in hunting through Ptolemy's catalogue of localities for clues to the sources out of which it was put together, the prehistory of the map is certainly too complex to be reconstructible in its entirety from the evidence at our disposal. We can, however, make plausible guesses about some of the sources that Marinos and Ptolemy would have found useful.

These, of course, would have included older maps, for just as Ptolemy used Marinos' world map as a basis for his revision, Marinos surely also consulted the maps of his predecessors (who, with the exception of Eratosthenes, are unknown to us). In many cases this would not have been a straightforward process of copying or reading off locations, because not all detailed maps would have been constructed according to a strict projection defining each locality's position on the globe. Some of the spatial distortions in Ptolemy's map might have arisen because Marinos was taking information from a map that represented shapes, distances, and directions schematically or qualitatively.

Whether directly or at some remove, the localities eventually inscribed in Marinos' or Ptolemy's map would have originally been recorded in some kind of text. Few if any of these textual resources would have been composed in the first instance with the cartographer in mind, so that information of critical importance for drawing a map was typically left out or very poorly supplied. Among the most helpful would have been the so-called *itineraria* and *periploi*, comprising verbal records of the sequence of and distances between places along roads and coasts, respectively.[28] Examples of both types of text survive from antiquity, although none of the extant documents can be demonstrated to have been among the specific sources for Ptolemy's map.

An *itinerarium* would typically provide lists of localities along a network of roads, with the intervening distances, as in the following excerpt from the *Itinerarium provinciarum Antonini Augusti*, a work compiled in the third century A.D.:[29]

| From Treveri to Agrippina, 78 leagues,[30] as follows: | |
|---|---|
| Beda village | 12 leagues |
| Ausava village | 12 leagues |
| Egorigium village | 12 leagues |
| Marcomagus village | 8 leagues |
| Belgica | 8 leagues |
| Tolbiacum village of the Sopeni | 10 leagues |
| Agrippina city | 16 leagues |
| From Treveri to Argentoratum, 128 miles: | |
| Baudobrica | 18 miles |
| Salisio | 22 miles |
| Vingium | 23 miles |
| Mogontiacum | 12 miles |
| Bormitomagus | 16 miles |
| Noviomagus | 18 miles |
| Argentoratum | 19 miles |

Lists like this would have supplied the cartographer with series of place names to be strung out at the appropriate distances; but one would have had to learn the general direction of each route from other sources, while the direction of each single stage would have been a matter of guesswork.

The most detailed *itineraria* available to Marinos and Ptolemy would probably have pertained to the Roman road system, but there were comparable texts

[28]Dilke 1985, 112–144.

[29]Cuntz 1929, 57; see Dilke 1985, 125.

[30]The *league* was a Gallic unit equivalent to 1.5 Roman miles.

also for regions outside the Roman Empire, such as the so-called *Parthian Stations* of Isidoros of Charax (first century A.D.), an *itinerarium* of the roads through the Parthian Empire that constituted part of the overland trade route from the Mediterranean to central Asia.[31] It was apparently only when journeys did not follow well-established roads that writers provided rough indications of directions as well as distances, as was the case with the African expeditions of Septimius Flaccus and Julius Maternus (1.8 and 10), and the trade route through northwestern China from the Stone Tower to Sēra described by Maes Titianus (1.11).

The *periplus* was a handbook, analogous to the *itinerarium*, but listing places and distances of maritime travel. Since sailing routes in antiquity usually followed coasts, seldom crossing open water, most of the surviving *periploi* provide the reader only with distances (usually in stades, sometimes in days of sail), not directions. The cartographer working from a *periplus* would therefore have to draw on other sources as well as his imagination to turn the list of places into a graphical outline, naturally taking advantage of indications of capes and bays to give some verisimilitude to the shape of the coast.

A surviving *periplus* that is of particular interest for Ptolemy's *Geography* is the *Periplus of the Erythraean Sea*, a guide for merchants to the trade routes along the African and south Asian coasts of the Red Sea and Indian Ocean.[32] This anonymous work was written about the middle of the first century A.D., and seems to have been comparable in character to sources of information concerning these regions that were available to Marinos and Ptolemy. Distances are specified in the *Periplus of the Erythraean Sea* sometimes in stades, but also often in "runs" or days of sail, which the cartographer would have to convert to stades using a conventional estimate of the distance sailed in a day (cf. *Geography* 1.9). On the other hand, it does give many indications of directions of sail, albeit rather imprecise ones (only the four cardinal directions are named). A clear sign that Ptolemy used a source resembling the *Periplus of the Erythraean Sea* is his inclusion, contrary to his usual practice, of remarks concerning local products and articles of trade in the part of the geographical catalogue delineating the coasts of the Indian Ocean beyond the Ganges and the island of Taprobanē (7.2–4).[33]

From the narrative writings of travelers and historians one might also have extracted place names and descriptions of physical features, but few precise indications of their geographical location. It is difficult to say how widely Marinos or Ptolemy surveyed this class of literature for data to incorporate in the map.

[31]Edited in Müller 1855–1861, 244–254; translation in Schoff 1914.

[32]Casson 1989. For the date of composition of the work, see his pp. 6–7.

[33]This is evidently what Ptolemy is alluding to in 2.1 where he writes that he may occasionally include "some bit of current knowledge [that] calls for a brief and worthwhile note."

One instance that has often been cited is a village in northern Germany, mentioned by no other classical author, that might have arisen from a misunderstanding of a sentence in the *Annals* of the Roman historian Tacitus.[34]

One can imagine Marinos or Ptolemy turning with relief from such materials to the handful of places for which they believed they had a satisfactory, astronomically determined latitude or longitude. Inaccurate and sparse though the astronomical data were, they provided the cartographer with his most satisfactory control of the broad outlines of the map: a loose framework of determined parallels and meridians between which one had to fit the otherwise hopelessly flexible strings of place names found in the other sources. Considering the indispensable role of this kind of data in the construction of the map, it should not be surprising that Marinos and Ptolemy admitted among them several traditional positions that would not have stood up to the scrutiny of careful observation.

The principle of measuring latitude by observing the sun's noon altitude on an equinox or solstice was already a commonplace in Eratosthenes' time; and Hipparchus evidently knew how to convert a given maximum length of daylight for a locality into its latitude.[35] Nevertheless Ptolemy writes (1.4) that "Hipparchus alone has transmitted… elevations of the north pole for a few cities… and [lists of] the [localities] that are situated on the same parallels." Hence so far as Ptolemy knew, the three centuries that had elapsed since Hipparchus had produced no significant advancement in collecting this kind of data.

Some of Hipparchus' latitudes for specific places can be recovered from Strabo and other sources.[36] For a few cities, such as Athens, Carthage, and Alexandria, Hipparchus had a ratio of a *gnōmōn* to its shadow on the equinox, which is simply tan $\phi$ where $\phi$ is the latitude. Others are assumed to be situated on the parallels associated with maximum lengths of daylight increasing by steps of a quarter-hour. When we compare these Hipparchian latitudes with the latitudes

[34]"Siatoutanda" (*Geography* 2.11), perhaps from the phrase, "The rebels having departed to ensure their safety [*ad sua tutanda*]" (Tacitus *Ann.* 4.73, Loeb 4.129). The resemblance (which was first noticed by H. Müller in 1837) may, however, be accidental; see Furneaux 1896, 1:11 n. 7. The *Annals* were not published before A.D. 116, which is after the latest datable contents of Ptolemy's map that can plausibly be ascribed to Marinos (Syme 1958 2:471–473).

[35]Strabo (2.1.18 and 2.5.34–43, Loeb 1:281–285 and 502–521) reports from Hipparchus a series of distances in stades between the parallels corresponding to various maximum lengths of daylight; as shown by Diller (1934), the numbers are in most cases accurate if one calculates assuming 700 stades per degree and assuming a value of 23°40' for the obliquity of the ecliptic (which is an accurate parameter, but not directly attested for Hipparchus). Dicks (1960, 192–194) and Neugebauer (1975a, 2:734 n. 14) criticize Diller's procedure; but Neugebauer's attempt (pp. 304–306) to explain the Hipparchian latitudes as generated by an arithmetical sequence accounts for fewer of the data.

[36]The table in Dicks 1960 (p. 193) lists these localities, but one must consult the original texts elsewhere in his volume to find how Hipparchus expressed their latitudes.

that Ptolemy assigns to the same places, we find that Ptolemy has often preserved Hipparchus' values, but not always.

Especially toward the northern and southern extremities of the map, when Hipparchus situated places on the parallels associated with particular maximum lengths of daylight, Ptolemy keeps them there. Meroē and Ptolemais Thērōn in Aithiopia south of Egypt are right on the parallel for which the longest day is 13 hours; Berenikē and Soēnē in Egypt are on the parallel for 13½ hours (which is also the Summer Tropic circle); Tyre in Phoenicia is on the parallel for 14¼ hours; Rhodes is on the parallel for 14½ hours; Byzantion in Thrace and Massalia in Gallia Narbonensia are on the parallel for 15¼ hours; and the mouths of the Borysthenēs are on the parallel for 16 hours. Among these latitudes, those for Ptolemais, Byzantion, and the mouths of the Borysthenēs are significantly in error, by as much as 2½°.[37] Another false latitude derived from a traditional value for the longest day is Ptolemy's placement of Babylon 35° north of the equator, which is 2½° too far north and corresponds to the assumption made by the ancient Babylonian astronomers that the ratio of longest to shortest day at Babylon is 3:2. In this instance, however, Ptolemy seems not to be following Hipparchus, who had situated Babylon at very nearly its correct latitude, 32°30'.[38]

One city for which Ptolemy retained a Hipparchian latitude derived from an equinoctial *gnōmōn* shadow is Alexandria at 31° (more precisely, 30°58', as we know from the *Almagest*). This is remarkably close to the truth, considering that it is obtained from a shadow ratio in small round numbers, 5:3. Still, it is surprising that Ptolemy did not detect from his own observations at Alexandria that the accurate latitude was about a quarter of a degree further north (31°13').

The only astronomical method available in antiquity of determining the interval in longitude between two places was to establish the difference in equinoctial hours between noon at the places in question by observing the local times of a lunar eclipse at both places. Ptolemy complains (1.4) that only a small number of records existed of eclipses seen at different places, and mentions a particular one "seen at Arbēla at the fifth hour and at Carthage at the second hour." This was the famous eclipse that took place on the evening of September 20, 331 B.C., eleven days before the battle of Gaugamela (near Arbēla, in Assyria) in which Alexander defeated Darius III of Persia. The association of the eclipse with this momentous event likely explains why sightings of it in different places lived in memory. The three-hour difference in the observed times for Arbēla and Carthage would amount to approximately 45° difference in longitude; and Ptolemy turns out to have actually assigned the two cities longitudes 45°10' apart.

[37]Ptolemy's erroneous latitude for Byzantion (later Constantinople) continued to be used there until the eleventh century, when better geographical data from Arabic sources became available.

[38]Dicks 1960, 134.

Ptolemy does not say that this was the *only* simultaneously observed eclipse available to him, although that may well have been the case.[39] But the example is revealing in ways that he could not have known. For observers at Arbēla the eclipse actually began about 1½ hours after sunset, was total from about 2⅔ hours to about 3¾ hours after sunset, and ended a little before 5 hours after sunset; at Carthage the times would have been about 2¼ hours earlier, so that observers there would have seen the moon already almost totally eclipsed when it rose, with totality beginning in the middle of the first hour of night and ending about the middle of the second hour. Thus Ptolemy's report is barely acceptable for Carthage if mid-eclipse is meant and "at the second hour" means the beginning of that hour, but in serious error for Arbēla whether it refers to the middle or the beginning of the eclipse. Remarkably, a second report exists of simultaneous observations of this same eclipse, but for a different pair of localities and with different times. According to the Roman writer Pliny the Elder (d. A.D. 79), the moon was eclipsed at the second hour of night at Arbēla, and at moonrise (i.e., sunset) at Syracuse.[40] Pliny's version accurately describes the times of the eclipse's beginning. [41]

Clearly the reports of eclipses could be inaccurate and inconsistent, especially when they derived from unscientific observation. But even if a report specified the correct hour and it was clear what stage of the eclipse was meant, a longitudinal difference deduced from the difference between times reported only by the hour within which the event took place would have been subject to errors of 15°, which makes the procedure practically worthless except as a control on the intervals between very widely separated places. Ptolemy's 45° longitudinal interval between Carthage and Arbēla is grossly in excess of the correct figure, which is close to 34°, but when expressed in terms of terrestrial units of distance the error almost vanishes, because his equivalent for one degree in stades is only about 82 percent of what it should be. Thus incorrect observational data combined with a defective estimate of the size of the earth led Ptolemy to a result that happened to be concordant with the basically accurate east-west intervals between places in the Mediterranean and the Near East that he took over from Marinos, which were probably obtained from the stade distances in *periploi* (1.12).

[39]Heron *Dioptra* 35 (Teubner ed., 3:302–307) demonstrates a method of determining the great-circle distance between two cities using a lunar eclipse that he says was observed at Alexandria and Rome, with a two-hour difference. Neugebauer (1938–1939) identified this as the eclipse of March 13, A.D. 62. Ptolemy's longitudes for Rome and Alexandria are 23°50' apart, rather than the 30° that would result from taking this eclipse report seriously.

[40]Pliny 2.180 (Loeb 1:313).

[41]Ginzel 1899, 184–185. The eclipse was also recorded in a contemporary Babylonian "Diary" tablet, but the times are unfortunately broken off; see Sachs and Hunger 1988–, 1:176–179.

## *Ptolemy's Map Projections and Coordinate Lists*

In introducing the principles of map-making, in *Geography* 1.20, Ptolemy refers to the two kinds of map, spherical and plane, and points out that although maps on spheres keep the earth's spherical shape and consequently preserve perfectly the relative proportions of intervals on the earth, they are usually too small to show all the things one wants to map, and they cannot be surveyed by the eye in a single glance. Plane maps, on the other hand, although they fulfill the two last demands, require "some method" to satisfy the first two.

Plane maps do not have to represent the spatial relationships between places in a definite, quantitative way. We know that world maps in classical antiquity could be highly schematic, like the circular map that appears in a group of medieval Greek astronomical manuscripts, in which, for example, Egypt and the upper Nile are portrayed as an oblique rectangle straddling a horizontal chord representing the Tropic of Cancer.[42] Herodotus (4.36, Loeb 2:235) and Aristotle (*Meteor.* 2.5 362b12, Loeb 181–183) both describe the world maps (*periodoi*) of their time as circular, with a ring-shaped Ocean entirely surrounding the land-mass of the *oikoumenē*; though the fact that both authors describe the plan of these maps as laughable shows that they had a conception that a map should somehow portray the regions of the world with roughly correct relative positions and sizes. The *Tabula Peutingeriana* ("Peutinger Table"), a medieval Latin map of the Roman Empire and its environs that is an indirect copy of a lost fourth century map, illustrates still another possibility: the map is a rectangular strip, nearly 7 meters wide but only 34 centimeters high, so that north-south distances are greatly compressed relative to east-west distances, and all outlines are accordingly distorted.[43] The lost source-map was designed primarily to exhibit the network of roads with their distances, for which there was no need to preserve much semblance to the shapes on the globe, and the map's dimensions were likely dictated by the original medium, possibly a papyrus roll.[44]

On the other hand, any world map that displayed localities in relation to a "graticule" (grid of principal parallels and meridians) would be practically forced to conform to a *projection*, that is, a mathematically definable rule for establishing a unique point on the planar surface corresponding to each point determined by a given parallel and meridian on the globe. And there is considerable evidence, especially in Strabo, that the "revision" of the traditional map of the world that Eratosthenes (c. 285–194 B.C.) presented in the third book of his

[42]Neugebauer 1975b (with illustration).

[43]A color reproduction of a section of the *Tabula Peutingeriana* is shown in *History of Cartography*, vol. 1, plate 5; for the whole, see Miller 1916 or Bosio 1983.

[44]Dilke 1985, 113–120.

*Geography* extensively employed geometrical constructions in relation to a grid of parallels and meridians.[45]

We may presume, therefore, that the history of map projections began not later than Eratosthenes in the third century B.C. Ptolemy tells us that Marinos criticized "absolutely all" previous methods of making plane maps, which implies that there had in fact been considerable experimentation with making such maps prior to his time. We know almost nothing about what these methods were, with the exception of the geographer Strabo's verbal description of a graticule suitable for the world map (early first century A.D.). Although frustratingly lacking in technical detail, the passage is worth quoting as the only surviving discussion of the topic before Ptolemy:[46]

> But [a world map] requires a large globe, so that the aforesaid segment of it [containing the *oikoumenē*], being such a small fraction of it, will be sufficient to hold the suitable parts of the *oikoumenē* with clarity and give an appropriate display to the spectators. Now if one can fashion [a globe] this large, it is better to do it in this way; and let it have a diameter not less than ten feet. But if one cannot make [a globe] of this size or not much smaller, one ought to draw [the map] on a planar surface of at least seven feet. For it will make little difference if instead of the circles, i.e., the parallels and meridians with which we show the *klimata* and directions and other variations and placements of the parts of the earth relative to each other and to the heavens, we draw straight lines, with parallel lines for the parallels, and perpendicular lines for the [meridians] perpendicular to them. [This is permissible] because the intellect is able easily to transfer the shape and size seen by the sight on a planar surface to the [imagined] curved and spherical [surface]. The same will apply to oblique circles [on the globe] and straight lines [corresponding to them on the map]. And though it is true that the meridians everywhere, since they are all described through the pole, all converge to one point on the globe, nevertheless it will not matter if on the planar surface one makes the straight lines for the meridians bend together only a little. For even this is not necessary in many situations when the lines [representing the meridians and parallels on the globe] are transferred to the planar surface and drawn as straight lines, nor is the convergence [of the meridians] as conspicuous as the curvature [of the globe].

Strabo evidently has in mind two ways of drawing the lines representing the circles of latitude and longitude. In the first, parallels of latitude are represented by horizontal straight lines, and meridians by vertical straight lines, so

[45]Strabo 2.1 (Loeb 1:253–361).

[46]Strabo 2.5.10 (Loeb 1:449–451).

that every parallel intersects every meridian exactly at right angles and the meridians, being represented by parallel lines, do not converge at all toward the north. In the second, the parallels of latitude are again drawn as horizontal lines, but the meridians converge a little at the north end of the map. This might mean that the meridians are drawn as straight lines inclined slightly from the vertical as if to meet at a point somewhere above the north end of the map, in which case they cannot all be perpendicular to the parallels. But Strabo may merely have in mind a slight inward curvature of the meridians only at the very top of the map, as if to suggest schematically their ultimate convergence while keeping them otherwise perpendicular to the equator and parallel to each other.[47]

Strabo appraises these representations only from the point of view of their adequacy in giving the general visual impression of the *oikoumenē* as it would be seen on a globe, and so he says nothing about what metrical properties of the map on the globe, such as distance, area, or direction, are preserved in either planar projection. At least in the version of the map with the meridians drawn throughout as parallels, one would presumably have kept the horizontal intervals between meridians on the map in correct proportion to the longitudinal intervals between the actual meridians, and likewise the vertical intervals between parallels on the map proportional to the latitudinal intervals between the actual parallels. Strabo's first grid would therefore have resulted, in modern terminology, in an *equirectangular cylindrical* projection, in which distances measured along all meridians would be portrayed in correct ratio to distances measured along, at most one parallel north of the equator, or along the equator itself. If he intended the meridians in the second version of the grid to be convergent straight lines, the projection would have resembled the so-called *Donis* or *trapezoidal* projection invented by Nicolaus Germanus in the 1460s, in which the meridians are drawn as straight lines converging so that distances measured along the top and bottom parallels are in correct ratio to each other, and only the central meridian and central parallel are at right angles and in correct proportionality of distances to each other.[48]

Whatever the variety of projections Marinos had to choose from, he had, according to Ptolemy, adopted just that mapping which was least successful in preserving proportionality of distances. In Marinos' map graticule (Fig. 10), the parallels of latitude are represented by a set of parallel straight lines and the meridians by another set of parallel straight lines at right angles to them, as in Strabo's projection. But Marinos also specified how distances along the paral-

[47]This is definitely what Strabo has in mind when he applies the same vocabulary to the courses of the Rhine and the Pyrenees in 4.5.1 (Loeb 2.253).

[48]"Donis" is a misreading of *donnus* or *dominus*, prefixed as an honorific to Nicolaus' name, apparently because he was a Benedictine.

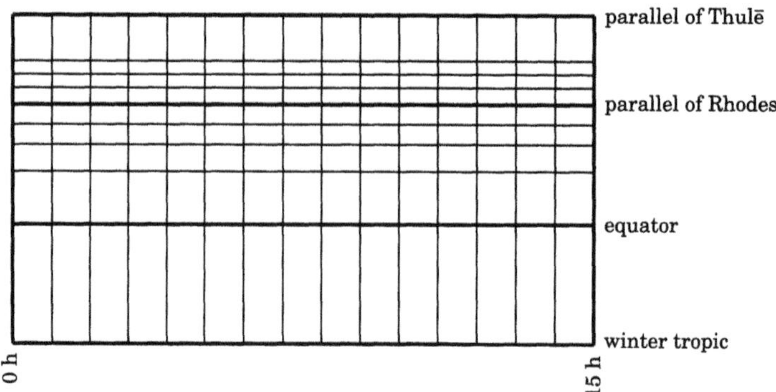


FIG. 10. Graticule of Marinos' projection

lels and meridians were to be represented in the projection. The ratio of the spacing of the lines separated by a given number of degrees of latitude relative to that of the lines separated by the same number of degrees of longitude was chosen to be 5:4, so that the ratio of a segment representing a degree in the east-west direction anywhere on the map to a segment representing a degree in the north-south direction is what it is on the globe at the latitude of Rhodes. As Ptolemy points out in 1.20, this means that the east-west spacing of places situated north or south of the parallel of Rhodes is progressively contracted the further south of Rhodes they are and progressively expanded the further north they are. The distortion would have been more pronounced in Marinos' map than in the map envisioned by Strabo, because Marinos' *oikoumenē* reaches significantly closer to the north pole than Strabo's does, and also extends to the equator and beyond, whereas Strabo assumed that the *oikoumenē* came to an end well north of the equator.

Ptolemy concedes this much to Marinos' choice of mapping, that if one imagines one's eye placed so that "the line of sight [is] directed at the middle of the northern quadrant of the sphere, in which most of the *oikoumenē* is mapped," and if the sphere is then revolved around its axis, each meridian in turn does appear as a straight line "when its plane falls through the apex of the sight." Hence, as a composite of a series of views of the sphere, the use of straight lines for meridians can be justified. But he also observes that to such an eye looking at the sphere, "the parallels... clearly give an appearance of circular segments bulging to the south," and a given pair of meridians "always cut off similar but unequal arcs on the parallels of different sizes, and always greater [arcs] on those nearer the equator."[49] Marinos' choice of projection lacks these properties.

[49]Not all of these statements are to be taken as a literal description of what is in fact seen, since in reality the parallels are seen as elliptical segments, not circular, and the portions of the arcs of different parallels between two meridians are not similar.

*Ptolemy's First Map*

After some further discussion, Ptolemy introduces the layout for his first map, in which he follows each statement about the geometry of the configuration with remarks about its effect (1.21):

> *First geometric criterion:* "It would be well to keep the lines representing the meridians straight, but [to have] those that represent the parallels as circular segments described about one and the same center, from which (imagined as the north pole) one will have to draw the meridian lines."
>
> *Its effect:* "Above all, the semblance of the spherical surface will be retained... with the meridians still remaining untilted with respect to the parallels [i.e., perpendicular to them] and still intersecting at that common pole."

> *Second geometric criterion:* "Since it is impossible to preserve for all the parallels their proportionality on the sphere, it would be adequate to keep this [proportionality] for the parallel through Thulē and the equator."
>
> *Its effect:* "The sides that enclose our [*oikoumenē*'s] latitudinal dimension [i.e., the bounding circular arcs representing the parallels of Thulē and the equator] will be in proper proportion to their true magnitudes."

> *Third geometric criterion:* "Divide [the parallel] that is to be drawn through Rhodes... in proportion to the meridian, that is in the approximate ratio of similar arcs of 5:4."
>
> *Its effect:* "The more familiar longitudinal dimension of the *oikoumenē* is in proper proportion to the latitudinal dimension."

The map that he produces has the following features:

1. The parallel bounding the *oikoumenē* on the north (the parallel through Thulē) is correctly represented relative to the size of the equator, i.e., the relative sizes of semicircles of the greatest and smallest parallels are correctly portrayed.
2. The longitudinal extent of the *oikoumenē* along the parallel of Rhodes relative to its latitudinal extent along the central meridian is faithfully represented on the map.
3. Each unit of distance along the straight lines representing the meridians between the parallels through anti-Meroē and Thulē faithfully rep-

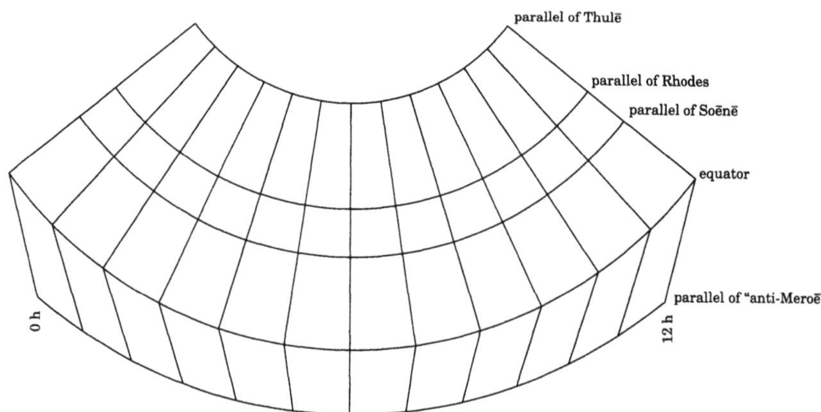


FIG. 11. Graticule of Ptolemy's first projection

> resents one degree of arc on the corresponding meridians on the globe. This is sometimes phrased as "distances are preserved on all radii."[50]

In modern nomenclature, this is a version of the *simple conical* projection (conical projections in general are those in which parallels are represented by concentric circles, and meridians by straight lines intersecting at a single point). This projection has the property that east-west distances are portrayed in correct proportionality to north-south distances only along the selected parallel through Rhodes, and are progressively exaggerated the further north or south one goes from this parallel. In Ptolemy's view, the distortion becomes intolerable for parallels south of the equator, because from this point on the actual parallels on the globe diminish in circumference, while the arcs representing them on the map continue to increase in length.

To compensate for this unwanted effect, Ptolemy introduces an ad hoc modification of his graticule. The arc representing the southernmost parallel to be included in the map is shortened to make it equal in arc length to the arc standing for the parallel that is the same distance north of the equator, and east-west distances along the two parallels are therefore in correct proportion to each other (though not to the meridians). The graticule is completed by drawing the parts of the meridians between the equator and the southernmost parallel as straight lines joining equal longitudes on the corresponding arcs (Fig. 11). The part of the map south of the equator is thus in a *pseudoconical* projection. This adjustment introduces a practical difficulty for anyone drawing the map, since one can no longer use a swinging ruler pegged to the common inter-

[50] E.g., Neugebauer 1975a, 2:881. Note that if this proportionality of distances is continued beyond the upper limit of Ptolemy's map, the north pole will be represented not by point *H*, where the meridians all converge, but by an arc with radius 7 units from *H*.

section of the meridians to locate points south of the equator. It also compromises the mathematical consistency of the projection, but to castigate this as a fault is to impute to Ptolemy a concept of map projection that was not his own.[51]

*Ptolemy's Second Map*

In contrast to the first map, where the eye is thought of as passing over each meridian in turn, in the second map the eye and the globe remain fixed relative to each other. Ptolemy attempts to produce the impression of the meridians and parallels as they would be seen when the axis of the visual cone joins the eye to the center of the sphere and passes through the intersection of the central meridian and the central parallel of the *oikoumenē*,[52] the eye being sufficiently far away from the globe so that for all practical purposes it sees a hemisphere. Such an eye will perceive two semicircles of great circles as straight lines. One of these is the visible half of the central meridian, and the other is a great circle passing through the two poles of the central meridian and the city of Soēnē (chosen because it lies exactly on the Summer Tropic circle). On the other hand, the same eye will view (1) the other meridian circles as a series of arcs equally balanced on either side of the central meridian, like right and left parentheses but increasingly curved the farther they are from the central meridian, and (2) the visible portions of the parallel circles as a series of concentric circular arcs.

Ptolemy also wishes to do this in such a way that (1) the lengths of the arcs of the parallel circles represented have the correct ratio to each other, not just for the equator and the parallel through Thulē, as in his first map, but also "as very nearly as possible for the other" parallels, and that (2) his map preserves the ratio "of the total latitudinal dimension to the total longitudinal dimension... not only for the parallel drawn through Rhodes..., but [at least] roughly for absolutely all [the parallels]" (1.24).

To accomplish this, Ptolemy imagines the viewing eye as seeing the visible hemisphere as a circle, bisected by two perpendicular diameters, whose length he arbitrarily sets at 180 units (representing linearly the 180° of the semicircle in the direction of the eye). In terms of these units he establishes where the equator crosses the central meridian at 23⅚ units below the center (because the eye is supposed to be above Soēnē, which is at latitude 23⅚° north), and then where the center of the circle representing the equator will be. Since this will also be the center of the other parallel circles, he is now able to draw the circles on which the following parallels lie: that of "anti-Meroē" (marking the southern limit of the *oikoumenē*), that of Soēnē, and that of Thulē (at the north-

[51]Berggren 1991, 134–138.

[52]His central meridian cuts through the Persian Gulf, passes slightly to the west of Persepolis, and then heads northward through the Caspian Sea and Skythia. The central parallel of latitude is the parallel of Soēnē, in Lower Egypt.

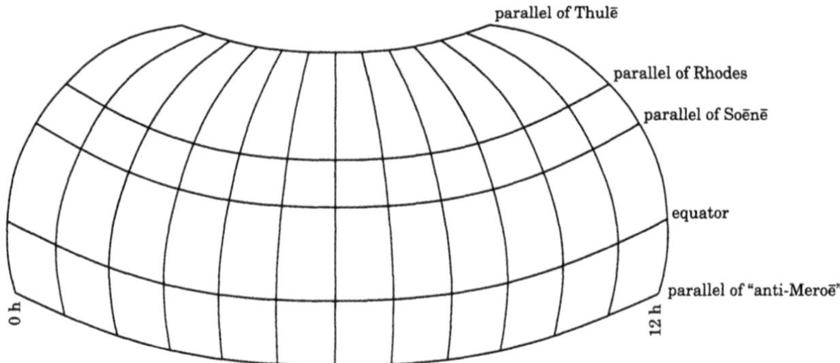


FIG. 12. Graticule of Ptolemy's second projection

ern limit). Taking arcs of five degrees to be equal to their chords, Ptolemy now marks off on each of these three parallels lengths corresponding to intervals of five degrees, and then joins triples of corresponding points with circular arcs to represent the meridians.

Figure 12 shows the resulting graticule. This is again a pseudoconical projection, since the parallels are drawn as concentric circular arcs, but the meridians are drawn as curves rather than as converging straight lines. Since a circular arc can be drawn through any three noncollinear points but not through any four, Ptolemy cannot keep distances measured along more than three of the parallels in exact proportionality with distances along the central meridian; and distances along the other meridians are distorted as a consequence of their curvature. If Ptolemy had made all the parallels proportionate in length to their actual lengths on the globe, and allowed the meridians to be drawn freely as the curves joining corresponding points on all parallels, he would have obtained the *Bonne* projection, which incidentally has the property of preserving areas of arbitrary regions on the globe.[53] Ptolemy, of course, would not have known this, and there is no suggestion that he, or any other ancient writer, had thought of preservation of areas as a desideratum in a map projection.

### *The Map in the Picture of the Ringed Globe*

In 7.6, Ptolemy sets out a long geometrical construction of an image of the terrestrial globe surrounded by rings representing the principal circles of the celestial sphere. The construction consists of two distinct parts: determining points

[53]Printed editions of the *Geography* of the late fifteenth century gave either Ptolemy's first projection (e.g., the Rome edition of 1490, reproduced in Nordenskiöld 1889, plate I) or the second (e.g., the Ulm edition of 1482). Bernardus Sylvanus (1511) and Johannes Werner (1514) were the first to generalize Ptolemy's second projection along the lines described above; see Neugebauer 1975a, 2:885–888.

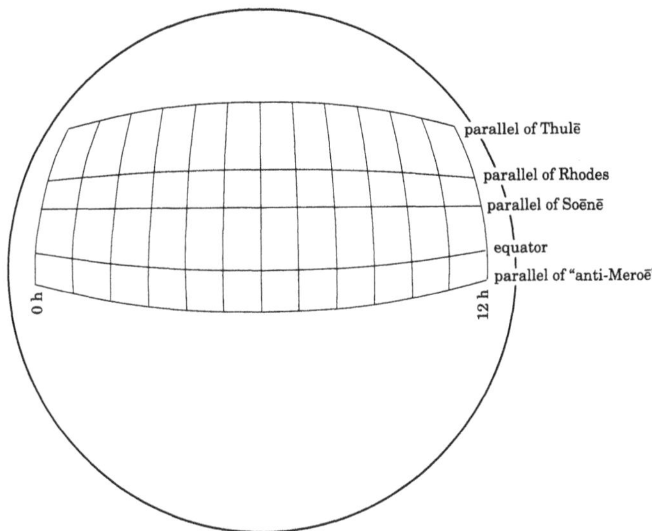


FIG. 13. Graticule of the projection in Ptolemy's picture of the ringed globe

through which the curves representing the various rings are to be drawn, and establishing a graticule for the map of the *oikoumenē* that is supposed to be visible between the rings. Ptolemy treats the two problems quite differently. The rings are drawn according to true linear perspective; that is, one imagines linear rays radiating from a point in space (representing the eye) through several points on each ring and onto a vertical plane, and the rings are drawn on that plane as ellipses passing through the projected points. This is in fact the unique example of a construction according to linear perspective surviving from antiquity. Ptolemy carries out the projection by treating the drawing plane first as the vertical plane containing the eye and the center of the globe, and thereafter as the plane of projection, which is at right angles to the former plane, so that the final drawing is made on the same plane as the geometrical construction of the perspective projection. This device, which eliminates the need for transferring measurements from one diagram to another, is reminiscent of the *analemma* constructions of sundial theory.[54]

Ptolemy's method for constructing the parallels of latitude portrayed on the terrestrial globe makes use of projected rays from the point representing the eye in a manner that superficially resembles the linear perspective used for the rings, but in fact the procedure has nothing to do with linear optics, and merely serves to generate a series of circular arcs that have their concavities facing a straight central parallel, thus qualitatively imitating the appearance of the actual parallel circles seen from an eyepoint in the plane of the chosen central parallel. The resulting projection (Fig. 13) resembles Ptolemy's second projec-

[54]Neugebauer 1975a, 2:839–856.

tion in using circular arcs to represent all meridians except the central one, which is a straight line, and in treating distances along this central meridian as proportional to the true distances on the globe; but the parallels are now laid out according to a plan analogous to that of the meridians instead of being drawn as concentric circular arcs. Again as in the second projection, proportionalities of distances are preserved along three parallels, along the top, bottom, and center of the map.

One may think of this third projection as a modification of Marinos' cylindrical projection, such that only one central meridian and one central parallel are drawn as straight lines in correct proportionality of distances, while the remaining parallels and meridians are drawn as circular arcs with curvature increasing as one goes further from the center of the map, to imitate the perspective appearance of the globe.

*The Regional Maps*

Having dismissed Marinos' cylindrical projection as unsuitable for a map of the entire *oikoumenē*, Ptolemy reintroduces it for the twenty-six regional maps into which he partitions the *oikoumenē* in Book 8. Each region is to be drawn in a graticule employing orthogonal straight lines for all meridians and parallels, and such that distances are represented in correct proportion along all meridians and along the central parallel for the region in question. The ratio of the lengths of one degree along the central parallel and along the meridians is stated in the caption to each regional map. Ptolemy left it to the cartographer, however, to find out just which meridians and parallels bound each region by finding the maximum and minimum longitudes and latitudes in the lists of coordinates. Someone after Ptolemy extracted these numbers, and listed them in a supplementary chapter that appears at the end of some manuscripts of the *Geography* (8.30 in Nobbe's edition).

*The Coordinate Lists*

Once the cartographer has constructed an appropriate graticule for the map of the world or one of the twenty-eight regions, the next task is to draw the map using the coordinate lists that make up the bulk of the *Geography* (Books 2.2–7.4). For this purpose, Ptolemy divided the *oikoumenē* into about eighty districts, which are grouped broadly into three continents (Europe, Libyē, and Asia), and within each continent are ordered loosely from west to east and from north to south. The chapters into which this part of the *Geography* is divided correspond to these districts.[55]

[55]It is not always obvious whether Ptolemy considers certain groupings of districts to belong together or not, so that the chapter divisions and the total number of districts are not definitely fixed.

Ptolemy refers to the districts in 2.1 as "provinces" and "satrapies," which would seem to identify them with the administrative divisions of the Roman and Parthian Empires, respectively. It does appear that Ptolemy (or Marinos) intended the districts contained by the Roman Empire to follow at least approximately the official borders of the provinces. On the other hand, Ptolemy's partition of Asia beyond the Roman frontier reflects the division of the Persian Empire into satrapies that was in effect in the time of Alexander the Great, and that had become part of the traditional apparatus of Greek geography. Ptolemy's map in fact makes little attempt to represent political geography, so that one cannot even tell from his map which districts belonged to the Roman Empire.

The map is composed of three kinds of object: one-dimensional (curvilinear) objects such as coastlines, the longer rivers, and some mountain ranges, which are to be drawn by connecting two or more points; pointlike objects such as cities, small islands, mountains, and the mouths of minor rivers; and peoples inhabiting small districts, who are located only in terms of the cities inside each district. Surprisingly, given that *itineraria* probably provided Marinos and Ptolemy with a good part of their geographical data, roads do not appear on the map. Each chapter begins with the definition of the outline of the province or satrapy, consisting of coastlines and borders (which sometimes coincide with rivers or mountain ranges). Borders that have already been described in a preceding chapter for the adjacent province are not repeated, and one generally has to refer back to the earlier chapter to find the last point from which drawing is to be continued. Cities and other features that lie on coasts are listed as part of the definition of the coastline. The coordinates defining the course of longer rivers are usually inserted as a digression at the point when the drawing of the coast has reached the river's mouth; for example, in 2.7 Ptolemy interrupts the description of the coast of Gallia Aquitanica when he has come to the mouth of the Garuna in order to insert the coordinates of two inland points that determine its course. The bends of some of the more complex rivers, such as the Nile, can only be drawn by inference after one has inscribed all the cities that are stated to be on one side or the other, a rare violation of Ptolemy's usual practice of giving specific longitudes and latitudes for all the cartographically significant points. Since the coordinates are specified only to the twelfth part of a degree, the resolution of the map is incapable of displaying distances smaller than that, so that, for example, the sizes and placements of offshore islands are not to scale.

## *The Manuscripts of the* Geography

Some understanding of the textual history of the *Geography* is indispensable for anyone who intends to study—or translate—the book from the available editions. We shall therefore give a brief outline of the ancestry of those manuscripts that are believed to be the most important for restoring Ptolemy's text.[56]

Our knowledge of the text of the *Geography* depends, for all practical purposes, on more than fifty Greek manuscripts, none older than the end of the thirteenth century. The genealogy of these manuscripts, though still not completely sorted out, is much better understood now than it was a century ago, when Nobbe and Müller published the last editions of the text to include the parts translated here.[57]

All manuscripts of the *Geography* seem to descend from a common ancestor later than Ptolemy. It is necessary to assume such an archetype (as opposed to two or more independent lines of descent from Ptolemy's autographs) in order to explain the errors common to all branches of the tradition, which are too numerous and often too serious to be the author's own.[58] Some variants between the manuscript families that are attributable to misreading the archetype indicate that it was written in capitals, which means that it was copied out earlier than the tenth century; it may in fact have dated back to late antiquity.[59]

The text of the *Geography* as it appeared in the archetype was already flawed, not only by accidents of copying, but also, apparently, by deliberate attempts to correct or improve Ptolemy's geographical data. One such instance, relating to part of the south coast of Italy, can be proved because some of Ptolemy's original coordinates have been handed down in the "Table of Noteworthy Cities" in his *Handy Tables*. Elsewhere Ptolemy's text may have been altered in ways that are more difficult to detect now.

The archetype also contained, perhaps at its end, some texts that did not originally belong to Ptolemy's work. The most important of these was a list of provinces classed according to Ptolemy's twenty-six regional maps, but with deviations from Ptolemy's provinces.[60] Among the other supplements are two

[56]Schnabel (1938, 5–33) describes forty-six Greek manuscripts containing all or part of the *Geography* (summarized on pp. 120–121 in a table and a rather tangled *stemma*). This list is not complete (see the addendum, p. 128, and Diller 1940b). A more up-to-date selective list is Diller 1966. We follow Diller's notations for the manuscripts, which are largely consistent with the conventions of Cuntz, Müller, and other previous editors of the *Geography*.

[57]The most significant contributions to the sorting out of the *Geography*'s manuscript tradition are Cuntz 1923 (esp. 1–37); Schnabel 1938; and Diller 1936, 1939, 1940a, 1940b, 1941, 1943, and 1966. Schnabel's monograph is still the most comprehensive treatment of the problem, but he unfortunately omitted much of the detailed argumentation behind his conclusions, some of which has since been shown to be incorrect. Fischer 1932a, although primarily concerned with the maps in the manuscripts, contains much that impinges on the history of the text.

[58]Examples are given by Cuntz (1923, 15). For some common errors in the description of Gaul, see pp. 126–127 (southern end of border between Gallia Aquitanica and Narbonensia; source of Sequana; Cemmena Mountains).

[59]Diller 1939, 229.

[60]Diller 1939, 93–95.

passages intended to be added to the captions for Ptolemy's world map and his picture of the globe (these may be by Ptolemy but are more probably spurious)[61], a small table that has to do with the motion of the sun north and south of the ecliptic,[62] a short poem in hexameters intended to be inscribed on a world map, and a note by a certain Agathos Daimon or Agathodaimon, engineer (*mēchanikos*) of Alexandria, announcing that he "sketched" (*hypetypōsa*) a map of the whole *oikoumenē* on the basis of the eight books of Ptolemy's *Geography*.[63]

Through most of the Middle Ages, Ptolemy's *Geography* was a rare and little-read text, a situation paralleled in the history of other ancient scientific and technical works.[64] The fortunes of the *Geography* changed abruptly around the year 1300, when several copies of the work—the earliest that survive—were made. From this time forth, manuscripts of the *Geography* proliferated.

The explanation of the *Geography*'s renewed popularity is likely to be found in the claim of the Byzantine scholar Maximos Planudes (c. 1255–1305) that he had "discovered through many toils the *geōgraphia* of Ptolemy, which had disappeared for many years."[65] We shall return in the following section to the question of what Planudes actually claims to have done, in particular whether he means that he rediscovered the text of the *Geography*, with or without maps, or that he reconstructed the maps from the text. But whatever the nature of Planudes' activity, there is a probable case for connecting it with a family of manuscripts of the *Geography* that date from about 1300.[66] These manuscripts were all copied, directly or indirectly, from a single lost copy. Three of the most important are beautiful large-format parchment codices containing maps. The cost of materials and workmanship must have been enormous, suggesting that these were presentation copies for very wealthy (or imperial) patrons.

The manuscripts of this family present a distinct recension of the text of the *Geography* characterized by extensive corrections of perceived errors in the text. Other alterations seem to be connected with the drawing of the maps, or to result from comparison with other authors. Such emendations are obviously a scholar's work.[67] We will refer to this version below as the "Byzantine revision."

These are the most important manuscripts of this group:

**U** = *Urbinas gr.* 82 (Vatican), a large parchment codex, c. A.D. 1300. The world map (employing Ptolemy's first projection in 1.24) follows the end of

[61]See p. 108 n. 1 for the former passage.

[62]Schnabel 1938, 64–67. The table is relevant to the problem of determining when the sun is directly overhead in tropical localities; see Neugebauer 1975a 2:936.

[63]Schnabel 1938, 92–94.

[64]See pp. 50–51.

[65]Kugéas 1909, 115–118.

[66]Diller 1940a. See also Diller 1936, 236–238; and Wilson 1981.

[67]For comparable examples in Planudes' studies of classical texts, see Wilson 1983, 232–236.

Book 7, while the twenty-six regional maps alternate with their respective captions in Book 8. A facsimile of this manuscript has been published.[68]

**K** = *Seragliensis gr.* 57 (Istanbul), a large parchment codex in the same format as **U**, c. A.D. 1300.

**F** = *Fabricianus gr.* 23 (Copenhagen), a single parchment sheet containing maps and text from Book 8, c. A.D. 1300. The manuscript to which this originally belonged was very like **K**.

**N** = *Bodl.* 3376, formerly *Selden.* 41 (Oxford), a paper manuscript containing the text of the *Geography* without maps, c. A.D. 1300.

**R** = *Marc. gr.* 516 (Venice), **V** = *Vat. gr.* 177 (Vatican), **W** = *Vat. gr.* 178, **C** = *Par. suppl. gr.* 119 (Paris). These fourteenth-century paper manuscripts are copies of a lost sister manuscript to **UKFN**. **R** contains a somewhat defective set of regional maps; **VWC** have none.

As has already been mentioned, the text of the *Geography* in this family shows clear signs of having undergone deliberate changes, which become apparent through comparison with other manuscripts to be described below. The redactor has here and there attempted to correct or smooth over difficulties in the sense and harshnesses in the language, often detecting real corruptions in the received text, but sometimes, one suspects, correcting Ptolemy himself. The spelling of many place names and some of the coordinates have been altered, evidently to resolve inconsistencies that became apparent in drawing maps. The captions of the regional maps in Book 8 have undergone fairly extensive revision.

We turn now to manuscripts that are partly or entirely free of the Byzantine revision:

**X** = *Vat. gr.* 191 (Vatican), a paper codex containing a large corpus of mathematical and scientific writings, copied by numerous hands and assembled about 1296.[69] The text of the *Geography* was originally copied by three hands, and some missing pages in the beginning have been replaced by a fourth; there are no maps. For some unknown reason the second and third scribes omitted all the numerical coordinates in the geographical catalogue from Book 5.13 on. In spite of this serious defect, **X** is a manuscript of the greatest importance for the text of the *Geography*, because it is the only copy that is uninfluenced by the Byzantine revision.

[68] Fischer 1932b.

[69] For the date and composition of the manuscript, see Turyn 1964, 89–97.

**Z** = *Pal. gr.* 314 (Vatican), a paper manuscript copied in about 1470. The text of this manuscript appears to derive from a text originally like that of **X** but extensively corrected against a manuscript related to **RVWC**.

**T** = *Burney* 111 (London), a fourteenth-century parchment manuscript with maps, apparently descended from a copy of a manuscript in which the Byzantine recension had been collated throughout against a manuscript of the unrevised text.

## *The Maps in the Manuscripts*

We have argued that, taken as a whole, the *Geography* is a unified composition that may be ascribed with confidence to its traditional author, Ptolemy. The same cannot be said, however, of the maps that accompany the work in many manuscripts. These are of two types: world maps, showing the whole of Ptolemy's *oikoumenē*, and regional maps. Practically all manuscripts containing maps have the world map, laid out according to Ptolemy's first map projection (except for **K**, which employs the second projection); but they fall into two different classes according to the number of regional maps they contain.

One of these classes, the manuscripts of so-called A version, contain twenty-six regional maps. These correspond closely to the maps to which Ptolemy refers in Book 8 in the following words: "We have made ten maps of Europe, four maps of Libyē, and twelve maps of the whole of Asia" (8.2). In these manuscripts the regional maps appear in Book 8 alternating with the relevant captions in Ptolemy's text. The manuscripts of the so-called B version contain sixty-four maps that portray smaller regions of the *oikoumenē* than Ptolemy's twenty-six regions; these are scattered at appropriate places in the catalogue of localities in Books 2–7.[70] In the Greek copies of both versions, the regional maps follow the cylindrical projection prescribed by Ptolemy, so that east-west distances along the central latitude of the map are in correct ratio to north-south distances, and the frame of each map is rectangular. Some later Latin copies adopt Nicolaus Germanus' refinement in which the meridians are drawn as converging straight lines in a trapezoidal frame, so that east-west distances at the top and bottom of the map are in true proportion to north-south distances.

The maps in most manuscripts of the A version are direct or indirect copies of those in **U**; this is obvious from the way that they reproduce trivial features such as the fictitious wiggles and bumps along the coast of unknown land to the south of the Indian Ocean on the world map, which are not derived from Ptolemy's coordinates. It also seems likely that the B version maps were produced from

[70]A few manuscripts of the B version also include four maps portraying the continents of Europe, Libya, and northern and southern Asia.

those of the A version for the convenience of those wanting to fit the *Geography* in manuscripts with smaller page dimensions, which necessitated more maps representing smaller areas. But were the maps in **U** and its sister manuscripts **K** and **F** themselves copies made by eye from maps in a lost manuscript, descending from an unbroken lineage beginning in antiquity, even from Ptolemy himself? Or were they reconstructed by Planudes or some other scholar about A.D. 1300 from the coordinates in the text, following Ptolemy's instructions? These questions have been the subject of much controversy in the past century. Without entering into a detailed discussion of the often complex arguments that have been presented on both sides, we can review some of the considerations that have led us to believe that, whatever the answers to the above questions may be, the maps that are present in the extant medieval copies are not an integral part of Ptolemy's work.

Although some scholars have gone so far as to doubt whether Ptolemy actually drew, or had drawn for him, the maps that he describes in the *Geography*, it seems hard to imagine how he could not have done so.[71] First of all, he could scarcely have compiled his lists of coordinates directly from Marinos' world map, because the places in Marinos' map had not only to be adjusted in accordance with Ptolemy's systematic reduction of the eastward and southward extensions of the *oikoumenē*, but also had to take account of the corrections and additions that existed in verbal form in Marinos' last publications and the reports of Ptolemy's own informants. As Ptolemy insists in 1.17, the way to detect and eliminate inconsistencies such as those he detects in Marinos' writings is to draw a map.

If we concede, as we surely must, that there were intermediate maps and sketches preceding the compilation of Ptolemy's geographical catalogue, it does not have to follow that Ptolemy incorporated actual maps in the manuscript of the *Geography* that he published. The first question in our minds must be whether it is plausible that Ptolemy would have presented such a comprehensive and well-thought-out plan of how to draw maps of the *oikoumenē* without actually trying it out. Certainly Ptolemy's description of the actual mechanics of map-making has the ring of something written by one who had actually made maps from coordinate lists. Statements such as the one (1.22) advising anyone preparing a map on a globe to make sure that the semicircular ruler swinging about the poles is "narrow in order not to obstruct many localities; and [to] let one of its edges pass precisely through the points [representing] the poles, so that we can use it to draw the meridians," or the remark (2.1) that he has arranged his catalogue of localities with a view to "convenience in the drawing of the map in every respect, namely progressing toward the right, with the hand

[71]Berger (1903, 640–641) and Bagrow (1946), among others, have denied that Ptolemy drew maps. See also Dilke 1985, 207 n. 28, for references to other advocates on either side of the question.

proceeding from the things that have already been inscribed to those that have not yet [been inscribed]," suggest that there is solid experience in map-making behind the presentation.[72]

But the maps may not have accompanied the text of the *Geography*, or even have been propagated in manuscript form at all. One mode of presenting a map of the world in antiquity—perhaps the most important one—was its erection in a public place. The most famous example of such a display was the map placed on the wall of a portico in Rome at the beginning of the first century A.D. by the emperor Augustus' friend Agrippa; and a similar map of the *oikoumenē* was apparently put up about A.D. 300 in a portico at Augustodunum (modern Autun in France).[73]

Now, if Ptolemy intended that his world map should be constructible directly from the catalogue of coordinates in the *Geography*, with all the localities visible and labeled, then he must have had a rather large map in mind, certainly no smaller than a meter in height and two in breadth.[74] In Ptolemy's time books such as the *Geography* were in the form of rolls of papyrus, which were commonly in the neighborhood of 30 centimeters, and very rarely as much as 60 centimeters, in height, which is about the height of the leaves in the medieval manuscripts of the A version. Thus a very tall papyrus roll would just have been able to accommodate the twenty-six regional maps of Book 8 in their full detail, but not the world map.[75]

The possibilities and limitations of maps in ancient manuscripts are no longer wholly a matter of conjecture now that a substantial fragment of a papyrus roll dating from approximately the middle of the first century B.C. and containing part of a Greek geographical treatise accompanied by a map has very recently come to knowledge.[76] The text in question is the description of Spain from the *Geography* of Artemidorus (c. 100 B.C.), a work known to Strabo, and the map, which follows (i.e., appears to the right of) the text. The height of the roll was (at least) 32.5 centimeters, which is exceptional for a literary roll of this

72Ptolemy's instructions for making the maps in the *Geography* are comparable to those describing the construction of observational instruments and the star globe in the *Almagest* and of experimental apparatus in his *Optics*.

73Dilke 1985, 41–54.

74Localities are placed on the map as close to each other as one-twelfth of a degree, and the rectangle containing the projection is roughly the equivalent of 90° from top to bottom and twice that in breadth.

75Diller (1939, 233 and 237) has shown that the medieval copies of the text of the *Geography* descend from a lost manuscript in which there were only about thirty-five lines to a page, which would have been too small for any of the maps.

76Gallazzi and Kramer 1998. The full contents of this remarkable papyrus, which is in a private collection, have not yet been published; our inferences based on the provisional description may in time have to be modified. The manuscript seems not to have been finished, so that the map lacks labels for the localities, and unused parts of the roll were subsequently used for artistic sketches.

period.[77] When complete, the map was more than 93.5 centimeters wide, so that the more or less equal north-south and east-west dimensions of the region must be portrayed with much lateral distortion, as in the *Tabula Peutingeriana*.

We have already mentioned the note by the Alexandrian engineer Agathos Daimon preserved in the manuscripts of the *Geography*, in which he states that he drew the world map according to Ptolemy's text. This note, which probably dates from antiquity (the man's name is unlikely to be Byzantine), could be imagined as the "signature" of a map drawn by Agathos Daimon in an ancestor of our manuscripts, although it does not in fact accompany the world map in those manuscripts that have one. Nothing, however, excludes the possibility that Agathos Daimon left the note as a testimonial to his success in applying Ptolemy's instructions to construct a map somewhere else.

Manuscript **X** has the following note following the end of Book 8, and written in the same hand as the preceding text: "Here he prescribes twenty-six charts; but in the actual map, twenty-seven. For he divides the tenth chart of Europe into two, putting Macedonia in one, and Epirus, Achaea, the Peloponnese, Crete, and Euboea in the other." These remarks apparently refer to Ptolemy's own maps, or maps that the writer takes to have been Ptolemy's. Again there is no suggestion that they were present in the manuscript in which the note was originally written. **X** contains no maps, and in the manuscripts that have the regional maps, the tenth map of Europe is not subdivided.

Maps based on the *Geography* are likely to have been seen by Pappus (fourth century A.D.) and al-Khwārizmī (eighth century), the authors of geographical works incorporating Ptolemaic data that will be discussed in the following section. In neither case need we presume that the maps accompanied Ptolemy's text. In the tenth century, the Arabic historian al-Mas'ūdī wrote in his *Fields of Gold* (ch. 8) that he had seen a book in Greek entitled *Geographia*, the author of which he refers to simply as "the philosopher," and in this manuscript were detailed descriptions of cities, mountains, seas, islands, and rivers.[78] Al-Mas'ūdī's "philosopher" has generally been taken to be Ptolemy, and since al-Mas'ūdī writes that the mountains and seas in the book were given various colors, it has also been inferred that it contained maps. But the account of what this book contained shows that it was definitely not Ptolemy's *Geography*: the numbers of features are all different, and Ptolemy did not say anything about the heights of mountains or the mines and precious stones in them. In another work al-Mas'ūdī refers to a *Geographia* purporting to be by Marinos that contained maps, which might have been the same book.[79]

[77]Gallazzi and Kramer 1998, 189.

[78]*Murūj al-dhahab wa ma'ādin al-jawhar*, ed. Barbier de Meynard and Pavet de Courteille, 183–185.

[79]See p. 23 note 24.

The more specific argument that the oldest extant Ptolemaic maps are products of the scholarly exertions of Maximos Planudes about A.D. 1300 depends primarily on a poem in hexameter that is entitled in one copy, "Heroic verses by the most wise monk Maximos Planudes on the *Geography* of Ptolemy, which had vanished for many years and then had been discovered by him through many toils."[80] The gist of the poem is as follows:

> What a great wonder, the way that Ptolemy has brought the whole world into view, just like someone making a map showing just a little city. I never saw anything so skillful, colorful, and elegant as this lovely *geōgraphia*. This work lay hidden for countless years and found no one to bring it to light. But the emperor Andronikos exhorted the bishop of Alexandria, who took great troubles that a certain free-spirited friend of the Byzantines should restore a likeness of the picture worthy of a king.

This can be interpreted in two ways. It has been taken as saying that Planudes had come across a manuscript of Ptolemy's *Geography*, which had fallen into oblivion, and that this old manuscript already contained the world map to which the opening lines of the poem clearly refer. But the poem as a whole, with its frequent allusions to the work involved in the rediscovery, is more likely to mean that he had taken great pains to rediscover the art of map-making set out in the treatise, and that the emperor Andronikos II had encouraged the patriarch of Alexandria, Athanasios II (who was in Constantinople at the time), to assume the patronage of the expensive project of reconstructing the map or maps. The word *geōgraphia* would mean not the book, but the map, as Ptolemy uses the word. This interpretation is supported by a second heading preceding Planudes' verses in another manuscript, which states that Planudes drew Ptolemy's map on the basis of Ptolemy's book and uninstructed by anyone else. Although neither poem nor titles mention that there was more than one map involved, it seems more plausible to assume, with Diller, that Planudes thought of the reconstruction of the world map and the twenty-six regional maps as a single exercise of *geōgraphia*, rather than that his exemplar had the regional maps and he restored just the world map.[81]

To sum up our conclusions from this evidence: There is no more reason to imagine that Ptolemy published his *Geography* in a form that incorporated the maps than there is to think that he provided a star globe along with the *Almagest*. The exceptionally large pages of such Byzantine copies as **U** and **K** are the minimum for the regional maps, and they are only able to hold the world map

[80]Stückelberger (1996) presents an edition and German translation of the whole poem with useful commentary, but arrives at conclusions different from those expressed here.

[81]Diller 1940a, 66.

because that map was drawn more or less freehand on the basis of the regional maps rather than directly from Ptolemy's coordinates. The transmission of Ptolemy's text certainly passed through a stage when the manuscripts were too small to contain the maps. Planudes and his assistants therefore probably had no pictorial models, and the success of their enterprise is proof that Ptolemy succeeded in his attempt to encode the map in words and numbers. The copies of the maps in later manuscripts and printed editions of the *Geography* were reproduced from Planudes' reconstructions.

## *Early Readers and Translators*

Ptolemy's *Geography* had other descendants besides the tradition of manuscripts in Greek. Writers starting with Ptolemy himself used the *Geography* for various purposes, extracting and preserving its contents in a new form. Nor was the work's heritage restricted to the Greek-speaking world. The early adaptations of the *Geography* are interesting as a record of the book's prolonged influence.

The earliest and most important adaptation of material from the *Geography* is Ptolemy's own list of important cities in his *Handy Tables*. As mentioned above, the *Handy Tables* is a set of astronomical tables, mostly extracted with modifications from the *Almagest*; it survives in several medieval copies.[82] The "List of Noteworthy Cities," which is found near the beginning of the *Handy Tables*, was certainly an original component of the work, since it is mentioned by Ptolemy in his brief introduction to the tables.[83] Transcriptions of the list have been published from two early (ninth century) copies.[84] The order of the cities, which is the same in the *Handy Tables* and in *Geography* 8, is determined by the plan of the *Geography*, which must therefore be the earlier work.[85]

About A.D. 300, the mathematician Pappus of Alexandria wrote a description of the *oikoumenē* that was based on the *Geography*. This work is known to us only through a medieval Armenian adaptation, although a few details survive through the process of abridgment and translation that are useful for studying the history of Ptolemy's work.[86] A scattering of references, sometimes of textual value, can be found in the late fourth-century Roman historian Ammianus Marcellinus, who used the *Geography* (perhaps through an intermediate adaptation) as a source especially for descriptions of the more distant parts of the

[82]The only edition, unsatisfactory by modern standards, is Halma 1822–1825.

[83]Greek text in Heiberg 1907, 161.

[84]Honigmann 1929, 193–224.

[85]We disagree with Schnabel's opinion (Schnabel 1930, 225–229) that the *Handy Tables* represents an earlier stage of Ptolemy's geographical researches. Schnabel's argues this from differences between the *Handy Tables* list and the *Geography* that are more plausibly to be ascribed to simple copying errors.

[86]Hewsen 1971 and 1992.

world.[87] The text of the *Geography* was also available in Italy in the middle of the sixth century, for it is mentioned by Cassiodorus (*Institutiones* 1.25): "you have the codex of Ptolemy, which sets out all localities so clearly that you would almost conclude that he was an inhabitant of all the regions."[88]

Much more dependent on Ptolemy than these authors, however, is another late writer, Marcianus of Heraclea (before the mid-sixth century),[89] who compiled a small handful of geographical works, very imperfectly preserved, of which the *Periplus of the Outer Sea* is the most important.[90] In this work, Marcianus gives a detailed account of the outline of the Asian part of the Indian Ocean and the European part of the Atlantic, following Ptolemy's account, and giving measurements of the distances between various points in stades, derived mathematically from Ptolemy's coordinates. Bits of the *Geography* are also quoted or adapted by two anonymous ninth-century Byzantine geographical compilations.[91] We next hear of the *Geography* in the *Chiliades* of Johannes Tzetzes (twelfth century), which incorporates a bizarre versification of excerpts from Books 3 and 5.[92] Then there is an interval of more than a century, during which there are no further traces of knowledge of the *Geography*, until we come to the efforts of Maximos Planudes described above.

The contents of the *Geography* were at least partially transmitted through translations and less direct means into the Arabic world as early as the ninth century.[93] However, no actual Arabic translation survives from the medieval period, and in the adaptations that we do possess, data from the *Geography* are mixed up with other sources.[94] Nevertheless, the presence in Arabic astronomical and geographical tables of many coordinates of longitude and latitude derived from the *Geography* has the potential for casting light on the history of the Greek text.[95] An important early Arabic adaptation of material from the *Geography* that has received particular attention is the geographical treatise

[87]Mommsen 1881; Polaschek 1965, cc. 764–772.

[88]There is no need to suppose, with Stückelberger (1996, 205), that Cassiodorus' manuscript contained maps.

[89]Diller 1952, 45.

[90]Edited in Müller (1855–61) v. 1, 515–76.

[91]Diller (1975) 38–41.

[92]*Chiliades* 11.888ff.

[93]According to Ibn al-Nadim in his *Fihrist* (Dodge 1970, 2:640), "Al-Kindi made a bad translation of it and then Thābit [b. Qurra (d. 901)] made an excellent Arabic translation. It is also extant in Syriac." Quite independent of these medieval translations—and with no significant effect on the medieval Islamic tradition of geography—a translation into Arabic was done shortly after the Turkish conquest of Constantinople in 1453, on the order of its conqueror, Sultan Mehmet II, by Georgios Amirutzes and his son. This translation is extant.

[94]For brief discussions of the *Geography*'s Arabic heritage, see Honigmann 1929, 112–122; and Dilke 1985, 155–157.

[95]Kennedy 1987.

*Kitāb ṣūrat al-'arḍ* ("Book of the picture of the world") ascribed to Abū Ja'far al-Khwārizmī (usually assumed to be the well-known ninth-century mathematician and astronomer Muḥammad Ibn Mūsā al-Khwārizmī), which describes a world map based ultimately on Ptolemy, but only through intermediaries—perhaps in Syriac—in the form of both text and map.[96]

The Renaissance Latin versions of the *Geography* depend entirely on the Byzantine Greek tradition and have little or no independent value for reconstructing Ptolemy's text. The first translation, by Jacopo d'Angelo, was finished in 1406 and was based on a composite text derived from two manuscripts.[97] This was printed many times from 1475 on, with cumulative revisions based sometimes on Greek manuscripts.[98]

## *Modern Editions and Translations of the* Geography

The Greek text of the *Geography* was first printed in 1533 at Basel, in an edition by Erasmus. This was followed over the next three centuries by several editions or partial editions.[99] The earliest that is still of much use, however, is that of F. W. Wilberg and C.H.F. Grashof, begun in 1838 and terminated prematurely in 1845 with Book 6.[100] This edition gives an apparatus reporting variant readings from several important manuscripts. For Book 6, this is the only critical text with apparatus.

The 1843–1845 edition of C.F.A. Nobbe, although it lacks apparatus and cites only a few manuscript variants (mostly in the spelling of place names) in an appendix, presents the most recent text of the entire *Geography*.[101] It is therefore necessary for the text from 7.5 to the end of the work, and it has useful indexes of place names and terminology.

Another edition was begun in 1883 by C. Müller, with a second (and final) volume published in 1901, after Müller's death, under the supervision of C.T. Fischer.[102] This edition presents the text, with apparatus and notes, only as far as the end of Book 5. Müller used numerous manuscripts, including most of those discussed above (excepting **U** and **K**). An inadequate classification of the manuscripts resulted in a cumbersome apparatus, citing readings from unimportant copies while omitting many important variants in the principal ones. Nevertheless, Müller's remains the best available critical text for Books 1 through 2.6, and 3.2 through 5.

[96]Mžik 1916 and 1926; and Wieber 1974.

[97]Diller 1966, ix–x.

[98]Codazzi 1948–1949.

[99]For a list, see Diller 1966.

[100]Wilberg 1838–1845.

[101]Nobbe 1843–1845.

[102]Müller 1883–1901.

In the 1920s appeared the two intentionally partial editions of Cuntz (2.7–3.1) and Renou (7.1–7.4).[103] Their greatest merit is in providing full citations of the variant readings in the best manuscripts representing the Byzantine recension as well as **Z** and **X**.

The *Geography* has seldom been translated into modern languages. Hans von Mžik's 1938 German translation of Book 1 and the first chapter of Book 2 was intended as the beginning of a complete translation of the work. No more was published, and this first part is not easy to come by. Mžik's rendering of Ptolemy is accurate, and it is accompanied by learned notes. The appendices by Friedrich Hopfner, dealing with technical topics including the map projections, are especially valuable. Germaine Aujac produced in 1993 a French version of parts of the *Geography* almost exactly coextensive with ours as part of a volume that also contains the geographical passages of the *Almagest* and *Tetrabiblos*. She provides a long introduction but comparatively light annotation. Like Mžik, Aujac chooses for the sake of smoothness a less strictly literal style of translation than we have preferred. Her interpretation of the text differs substantially from ours in about a dozen obscure passages.

The only previous English rendering of the *Geography* is that of Edward Luther Stevenson; it was originally published in 1932 in a very small edition, and was reprinted in 1991. Stevenson's translation covers nearly the entire work (omitting the regional map captions of Book 8) and is accompanied by black-and-white photographs of the maps from one of Nicolaus Germanus' manuscripts, the *codex Ebnerianus* of the New York Public Library. These are regrettably its only virtues. Stevenson appears to have based his version primarily, if not exclusively, on the Renaissance Latin texts of the *Geography*, and very frequently misunderstood even them. Diller's comment that "to speak the plain truth, there is not a single paragraph that does not betray some essential and often gross error" is no hyperbole.[104]

## *Our Translation*

Our translation uses as its base text the Greek text of Nobbe, which has the advantage of being complete and fairly accessible; however, we have also regularly consulted the incomplete editions of Wilberg and Müller and photographs of the manuscripts **U**, **K**, **N**, **T**, and **X**. For Books 7 and 8, where we had only Nobbe as a printed text, we established a provisional text on the basis of the manuscripts cited above. Whenever we have chosen a reading of the Greek text significantly different from that of Nobbe, we have reported this in Appendix G, "Textual Notes."

[103]Cuntz 1923; Renou 1925.
[104]Diller 1935, 536.

We have tried to be as faithful to Ptolemy's way of putting things as is consistent with a translation. In particular, this has meant that terms such as *oikoumenē* and *klimata*, the names of winds and units of measurement, and the names of most localities have simply been transliterated. (Where necessary, explanatory notes have been added.) We have not, however, carried this fidelity to the point of reproducing Ptolemy's often exceedingly long sentences, which we have not hesitated to break up.

Square brackets in the translation enclose small additions of our own, intended to make a smoother translation or to clarify ambiguities. Most of these additions are such as would be understood from the grammar by a reader of the Greek text, but some definitely reflect our interpretation of the text.

We have in a few cases translated place names when we judged that it would help the reader, as in the cases of the "Islands of the Blest," the "Caspian Gates," and the "Stone Tower." In a few other cases, we have used modern forms of the names when they diverge little from those in Ptolemy and when the site the modern form refers to is the same as that referred to by the ancient, as, for example, Rhodes, Carthage, or Smyrna. For most places, however, we have retained the ancient names. For those in the western part of Ptolemy's *oikoumenē*, where Ptolemy is transliterating Latin forms into Greek, we have used Latin forms. For the eastern part, we have transliterated the Greek names as Ptolemy gives them. Preserving the Greek spelling often helps the reader to avoid confusing ancient and modern geographical entities that have essentially the same name but refer to different places: for example, we have kept Ptolemy's "Aithiopia" as the name of the southernmost districts of Africa rather than modernizing to "Ethiopia," and "Libyē" as the name for the entire continent of Africa. The possibility of confusion could not be avoided in a few instances, however, such as the name "Africa" itself, by which Ptolemy means the Roman province centered on Carthage. In general, the reader should not take it for granted that names that exist on the modern map refer to the same places in Ptolemy. The Geographical Index (Appendix H) attempts where possible to give modern equivalents for the localities mentioned in the text.

For ancient authors whose works survive and for historical personages likely to be found in reference works, we have used the forms of their names that are standard in modern English scholarship. Other Greek personal names are transliterated.

PTOLEMY

# Guide to Drawing a Map of the World

# Book 1

## 1. On the difference between world cartography and regional cartography

World cartography[1] is an imitation through drawing of the entire known part of the world together with the things that are, broadly speaking, connected with it. It differs from regional cartography in that regional cartography, as an independent discipline, sets out the individual localities, each one independently and by itself, registering practically everything down to the least thing therein (for example, harbors, towns, districts, branches of principal rivers, and so on), while the essence of world cartography is to show the known world as a single and continuous entity, its nature and how it is situated, [taking account] only of the things that are associated with it in its broader, general outlines (such as gulfs, great cities, the more notable peoples and rivers, and the more noteworthy things of each kind).

The goal of regional cartography is an impression of a part, as when one makes an image of just an ear or an eye; but [the goal] of world cartography is a general view, analogous to making a portrait of the whole head. That is, whenever a portrait is to be made, one has to fit in the main parts [of the body] in a determined pattern and an order of priority. Furthermore the [surfaces] that are going to hold the drawings ought to be of a suitable size for the spacing of the visual rays[2] at an appropriate distance [from the spectator], whether the drawing be of whole or part, so that everything will be grasped by the sense [of sight].

[1]We thus translate *geōgraphia* in accordance with the restricted sense that Ptolemy defines for the word in this chapter. "Regional cartography" represents Ptolemy's *chōrographia*. Other Greek authors, such as Strabo, use *geōgraphia* to mean a written geographical work.

[2]What Ptolemy is asserting is that when making any picture, one should decide how big it should be in accordance with the level of detail and the expected distance of the spectator. He expresses the fact that the eye perceives less detail with greater distance in terms of the concept of visual rays found in Euclid's *Optics*. The rays were supposed to radiate from the eye to the object of vision and to transmit color to the eye. Euclid assumes that there are a finite number of rays separated by spaces that widen with increasing distance from the eye; the gaps explain loss of resolution when one views an object at a distance. In his own *Optics*, Ptolemy rejects these discrete rays in favor of a continuous visual cone that emanates from the eye. See Smith 1996, 91–92.

In the same way, reason and convenience would both seem to dictate that it should be the task of regional cartography to present together even the most minute features, while world cartography [should present] the countries themselves along with their grosser features. This is because with respect to the *oikoumenē*[3] it is the geographical placements of countries that are the main parts, [namely] the ones that are well placed and of suitable sizes [for a map], whereas the various things contained in these [countries have the same relationship] with respect to [the countries themselves].

Regional cartography deals above all with the qualities rather than the quantities of the things that it sets down; it attends everywhere to likeness, and not so much to proportional placements.[4] World cartography, on the other hand, [deals] with the quantities more than the qualities, since it gives consideration to the proportionality of distances for all things, but to likeness only as far as the coarser outlines [of the features], and only with respect to mere shape. Consequently, regional cartography requires landscape drawing, and no one but a man skilled in drawing would do regional cartography. But world cartography does not [require this] at all, since it enables one to show the positions and general configurations [of features] purely by means of lines and labels.

For these reasons, [regional cartography] has no need of mathematical[5] method, but here [in world cartography] this element takes absolute precedence. Thus the first thing that one has to investigate is the earth's shape, size, and position with respect to its surroundings [i.e., the heavens], so that it will be possible to speak of its known part, how large it is and what it is like, and moreover [so that it will be possible to specify] under which parallels of the celestial sphere each of the localities in this [known part] lies. From this last, one can also determine the lengths of nights and days, which stars reach the zenith or are always borne above or below the horizon,[6] and all the things that we associate with the subject of habitations.[7]

[3]Literally, "the inhabited [part of the world]." This technical term of Greek geography is sometimes used interchangeably with "the known part of the world," although the concepts are not strictly equivalent.

[4]This passage makes it clear that the "regional cartography" that Ptolemy has in mind not only covers smaller areas of the world than his world cartography, but also follows different principles. It seems to have been something closer to landscape drawing, incorporating lifelike images of features of the area portrayed. The closest counterparts we have from antiquity are mosaic maps such as the Madaba mosaic; see Dilke 1985, 148–153.

[5]Ptolemy uses the term "mathematics" not only for the abstract sciences of numbers and geometry but also for subjects such as optics, harmonics, and astronomy, in which physical objects are investigated from the point of view of their mathematical properties.

[6]Literally, "below the earth."

[7]"Habitations" (*oikēseis*) means the determination of the astronomical phenomena characteristic for particular terrestrial latitudes. Book 2 of the *Almagest* is largely devoted to a theoretical treatment of this topic.

These things belong to the loftiest and loveliest of intellectual pursuits, namely to exhibit to human understanding through mathematics [both] the heavens themselves in their physical nature (since they can be seen in their revolution about us), and [the nature of] the earth through a portrait (since the real [earth], being enormous and not surrounding us, cannot be inspected by any one person either as a whole or part by part).[8]

## 2. *On the prerequisites for world cartography*

We shall let this serve as a brief sketch of the purpose of anyone who would be a world cartographer, and how he differs from the regional cartographer. Our present object is to map our *oikoumenē* as far as possible in proportionality with the real [*oikoumenē*]. But at the outset we think it is necessary to state clearly that the first step in a proceeding of this kind is systematic research, assembling the maximum of knowledge from the reports of people with scientific training who have toured the individual countries; and that the inquiry and reporting is partly a matter of surveying, and partly of astronomical observation. The surveying component is that which indicates the relative positions of localities solely through measurement of distances; the astronomical component [is that which does the same] by means of the phenomena [obtained] from astronomical sighting and shadow-casting instruments.[9] Astronomical observation is a self-sufficient thing and less subject to error, while surveying is cruder and incomplete without [astronomical observation].

For, in the first place, in either procedure one has to assume as known the absolute direction of the interval between the two localities in question, since it is necessary to know not merely how far this [place] is from that, but also in which direction, that is, to the north, say, or to the east, or more refined directions than these. But one cannot find this out accurately without observation by means of the aforesaid instruments, from which the direction of the meridian line [with respect to one's horizon], and thereby [the absolute directions] of the traversed intervals, are easily demonstrated at any place and time.

[8]This rather obscure peroration entwines two ideas: that astronomy and geography are parts of a single rational science, and that whereas astronomy can make its demonstrations using the heavens themselves as a visible object of study, geography must make use of maps. We are inside the celestial sphere, and can behold half of it at once. By way of contrast, our position on the surface of the earth prevents us from taking in the earth's form at a glance, and it is too large for any single person to explore.

[9]A sighting instrument (*astrolabon*) is one that permits the direct measurement of the apparent position of a heavenly body through a diopter, for example, Ptolemy's armillary spheres (the *astrolabon* described by Ptolemy in *Almagest* 5.1 or the *meteōroskopeion* mentioned below in 1.3). A shadow-casting instrument could be a simple *gnōmōn* or upright stick, used to determine the sun's altitude.

In the next place, even when this [direction] has been given, having a measurement of distance in stades does not guarantee that the [interval] we find is the correct one, because one seldom encounters rectilinear journeys on account of the numerous diversions that are involved in both land and sea travel. For land journeys one has to estimate the surplus [in the reported distance] corresponding to the kind and magnitude of the diversions and subtract this from the total of stades to find the [number of stades] of the rectilinear [route]. For sea journeys one also has to account for the variation in speed corresponding to the blowing of the winds, since at least over long periods these do not maintain constant force. But even if the interval between the localities traveled through has been accurately determined, this does not also yield its ratio to the whole circumference of the earth, or its position with respect to the equator and poles.

The [method] using the [astronomical] phenomena determines each of these things accurately, since it shows the magnitudes of the arcs that the parallel and meridian circles drawn through the given localities cut off on each other—the arcs, that is, that the parallels [cut off on the meridians] between themselves and the equator, and [those that the meridians cut off] between themselves on the equator and on the parallels. [The astronomical method] also reveals the size of the arc that the two localities cut off along the great circle drawn through them on the earth. [This method] does not even need reckoning in stades, either to get the ratios of the earth's parts [with respect to each other and the whole], or in the entire process of map-making. It is enough to assume that [the earth's] circumference comprises any arbitrary number of units, and then to show how many [such units] make up the specific intervals along the great circles drawn on [the earth].

Admittedly, [the astronomical method] will not [also be able to yield] the division of the whole circumference or its parts into the established and familiar measures of length [used in] our distance measurements. For this sole reason it has been necessary to match a single rectilinear route [on the earth] to the [geometrically] similar great-circle arc on the surrounding [celestial sphere] and, having determined the ratio of this [arc] to the circle by means of the [astronomical] phenomena, and the number of stades in the route beneath it by means of distance measurement, to produce from the given part [of the circumference] the number of stades in the whole circumference. For it has already been mathematically determined that the continuous surface of land and water is (as regards its broad features) spherical and concentric with the celestial sphere [*Almagest* 1.4–5], so that every plane produced through the [common] center makes as its intersections with the aforesaid surfaces [of the terrestrial and celestial spheres] great circles on [the spheres], and angles in [this plane] at the center cut off similar arcs on the [celestial and terrestrial great] circles. As it happens, although the number of stades in intervals on the earth (if they are straight) can be determined from distance measurements, their ratio to the whole circumference cannot [be determined] at all from [distance measurements] because of the impossibility of making the comparison.[10] But [this ratio can be determined] from the similar arc of the circle on the surrounding [celestial sphere], because one can determine the ratio of this [similar arc] to the circumference [i.e., the great circle] to which it belongs, and this [ratio] is the same as that of the similar segment along [the surface of] the earth to the great circle on [the earth].

## 3. *How the number of stades in the earth's circumference can be obtained from the number of stades in an arbitrary rectilinear interval, and vice versa, even if [the interval] is not on a single meridian*

Now, our predecessors looked not just for a rectilinear interval on the earth to treat as an arc of a great circle, but also one that was directed in the plane of a single meridian.[11] Using shadow-casting instruments, they observed the zenith points at the two ends of the interval, and obtained directly the arc of the meridian cut off by [the zenith points], which was [geometrically] similar to [the arc] of the journey [between the two locations]. This is because these things were set up (as we said) in a single plane (since the lines produced through the [two] ends [of the journey] to the zenith points intersect), and because the point of intersection is the common center of the circles. Hence they assumed that the fraction that the arc between the zenith points was seen to be of the circle through the [celestial] poles [i.e., the common meridian of the two locations] was the same fraction that the interval on the earth was of the whole [earth's] circumference.

We, however, have established by means of the construction of a meteoroscopic instrument[12] that the [same] object can be achieved even if we

[10]Ptolemy means that because of the immensity of the terrestrial globe, one cannot directly measure its circumference or apprehend that a given measured distance is a particular fraction of the whole circuit.

[11]I.e., one place of observation was assumed to be due south of the other. This is true of Eratosthenes' famous measurement of the size of the earth based on the interval from Alexandria to Soēnē, as well as the similar method ascribed to Posidonius, based on the interval from Alexandria to Rhodes. See Neugebauer 1975a, 2:652–654; and Taisbak 1974.

[12]Ptolemy described his "meteoroscope" (*meteōroskopeion*) in a lost work known to us through references by Proclus and Pappus. It was an armillary sphere with nine rings, i.e., two more than the *astrolabon* of the *Almagest*. Ptolemy's armillary sphere had three rings for the ecliptic system, three for the equator system, and one sliding ring for sighting. From the present context it is clear that Ptolemy had added to his earlier instrument further rings for the horizon system. For an attempted reconstruction and discussion of how the instrument could have performed the tasks described in this chapter, see Rome 1927.

take the circle along the measured interval such that it is not through the poles,[13] but [is instead] any great circle, by observing in the same way the elevations [of the celestial pole] at the [two] endpoints [of the terrestrial interval] as well as the direction that the interval has with respect to one of the meridians [through the endpoints]. Using [the meteoroscope] we can easily obtain, among many other extremely useful things, the elevation of the north [celestial] pole at the place of observation on any day or night, and at any hour the direction of the meridian and [the directions] of routes with respect to [the meridian] (that is, the size of the angles that the great circle described through the route makes with the meridian at the zenith point).[14] With these [quantities known] we can show right on the meteoroscope the arc in question [of the great circle through the two locations] as well as the [arc] on the equator that the two meridians (if they are distinct) cut off.[15] Hence by this procedure the total number of stades of the [earth's] circumference can be found from just one rectilinear interval measured on the earth, and thereby also [the number of stades] of the other intervals without measuring the distances, even if they are throughout not rectilinear or along a single meridian or parallel, so long as the general trend of the direction and the elevations [of the celestial pole] at the endpoints have been carefully determined. This is because one can conversely compute the number of stades [of such an interval] easily from the established circumference of the whole [earth] using the ratio of the arc subtending the interval to the great circle.

## *4. That it is necessary to give priority to the [astronomical] phenomena over [data] from records of travel*

These things being so, if the people who visited the individual countries had happened to make use of some such observations, it would have been possible to make the map of the *oikoumenē* with absolutely no error.[16] But Hipparchus alone has transmitted to us [observed] elevations of the [celestial] north pole for a few cities, [i.e., few when] compared to the multitude of [cities] to be recorded in the world cartography, and [lists of] the [localities] that are situated on the same parallels.[17] And a few of those who came after him [have transmitted] some of the localities that are "oppositely situated"[18] (not [meaning] those

[13]That is, if we take two points of observation on the earth that are not on the same meridian.

[14]Actually, the angle between the meridian and the *celestial* great circle directly above the terrestrial interval is meant.

[15]See Textual Notes (Appendix G).

[16]On the astronomically determined positions available to Ptolemy and the use he made of them, see pp. 28–30.

[17]Literally "under the same parallels," the parallels being thought of as on the celestial sphere.

[18]Ptolemy uses *antikeisthai* ("to lie opposite") as a technical term for "to lie on one meridian."

that are equidistant from the equator, but simply those that are on a single meridian, based on the fact that one sails from one to another of them by *Aparktias* or *Notos* winds). Most intervals, however, and especially those to the east or west, have been reported in a cruder manner, not because those who undertook the researches were careless, but perhaps because it was not yet understood how useful the more mathematical mode of investigation is, and because no one bothered to record more lunar eclipses that were observed simultaneously at different localities (such as the one that was seen at Arbēla at the fifth hour and at Carthage at the second hour),[19] from which it would have been clear how many equinoctial time units separated the localities to the east or west. It would therefore also be reasonable for one who intended to practice world cartography following these [principles] to give priority in his map to the [features] that have been obtained through the more accurate observations, as foundations, so to speak, but to fit [the features] that come from the other [kinds of data] to these, until their positions with respect to each other and to the first [features] stand as much as possible in agreement with those reports that are less subject to error.

## *5. That it is necessary to follow the most recent researches because of changes in the world over time*

The foregoing would provide a plausible basis for the project of drawing a map. But in all subjects that have not reached a state of complete knowledge, whether because they are too vast, or because they do not always remain the same, the passage of time always makes far more accurate research possible; and such is the case with world cartography, too. For the consensus of the very reports that have been made at various times is that many parts of our *oikoumenē* have not reached our knowledge because its size has made them inaccessible, while other [parts] have been described falsely because of the carelessness of the people who undertook the researches; and some [parts] are themselves different now from what they were before because features have ceased to exist or have changed. Hence here [in world cartography], too, it is necessary to follow in general the latest reports that we possess, while being on guard for what is and is not plausible in both the exposition of current research and the criticism of earlier researches.

## *6. On Marinos' guide to world cartography*

Marinos of Tyre seems to be the latest [author] in our time to have undertaken this subject, and he has done it with absolute diligence. He has clearly laid his

[19]On this eclipse, see pp. 29–30.

hands on numerous records of research besides those that had come to knowledge still earlier, and treated those of nearly all his predecessors with care, giving appropriate correction to everything that he found that either they or he himself, at first, had trusted without good reason, as can be seen from his publications of the revision of the geographical map, which are numerous.

Now if we saw no defect in his final compilation, we would content ourselves with making the map of the *oikoumenē* on the basis of these writings alone, without taking any more trouble about it. Since, however, even he turns out to have given assent to certain things that have not been creditably established, and in many respects not to have given due thought to the method of map-making, with a view either to convenience or to the preservation of proportionality, we have justifiably been induced to contribute as much as we think necessary to the man's work to make it more logical and easier to use. We will do this as concisely as possible, starting with a brief examination of each kind of thing that needs some comment.

And the first of these is about the research on the basis of which he thinks the longitudinal dimension of the known world has to be increased to the east, and its latitudinal dimension to the south. For we can reasonably call the dimension of the surface in question [i.e., of the *oikoumenē*] from east to west "longitude," and that from north to south "latitude," since we call the [dimensions] that are parallel to these in the celestial motions by the same names, too, and because in general we use "longitude" [i.e., "length"] for the greater dimension, and it is agreed by absolutely everybody that the dimension from east to west of the *oikoumenē*, too, is much greater than that from north to south.

## *7. Revision of Marinos' latitudinal dimension of the known world on the basis of the [astronomical] phenomena*

First, in the case of the latitudinal dimension, [Marinos], too, assumes [as we do] that the island of Thulē is on the parallel that marks the most northerly limit of our known world and shows as best he can that this parallel is approximately 63° from the equator (where the meridian circle is 360°),[20] or 31,500 stades [from the equator] (assuming that a degree contains approximately 500 stades).[21] Then he sets the country of the Aithiopians called Agisymba, and Cape Prason, on the parallel that marks the southernmost limit of the known world, and he puts this at the Winter Tropic [circle]. Hence, according to him, the whole latitudinal dimension of the *oikoumenē* (adding the interval between the equa-

[20]Such definitions of the degree are frequent in Ptolemy and other ancient scientific authors, and allow for other definitions of the unit. The term *moira*, "degree," literally signifies merely a division or unit, and in some contexts we choose to translate it thus.

[21]See pp. 21–22.

tor and the Winter Tropic [to the 63°]) amounts to approximately 87°, or 43,500 stades. He tries to show that [his] southern limit is plausible both by certain [astronomical] phenomena (as he supposes them to be) and by the records of land and sea journeys. Each of these has to be checked for excess.

As to the phenomena, he says the following in the third [book of his] compilation (I quote): "For in the torrid zone the ecliptic passes overhead so that shadows alternate there,[22] and all the stars set and rise, except for Ursa Minor. The whole of [Ursa Minor] begins to be always visible when one is 500 stades north of Okēlis. This is because the parallel through Okēlis is elevated 11⅖° [north of the equator], and Hipparchus reports that the southernmost star of Ursa Minor (the last of the tail [α UMi]) is 12⅖° from the pole.[23] Moreover, when one is going from the equator toward the Summer Tropic, the north pole is always elevated above the horizon, while the south pole is below the horizon; but when one is proceeding from the equator toward the Winter Tropic, the south pole is raised above the horizon, while the north pole is below the horizon." Now in these words he is describing merely what *ought* to occur in locations on the equator or between the tropics; but he does not tell us whether there has actually been any research into the phenomena on [parallels] south of the equator, for example, that somewhere some stars that are south of the [celestial] equator reach the zenith, or that noon shadows point south at the equinoxes, or that *all* the stars of Ursa Minor rise or set, or even that some of them are not visible at all, whereas the south pole is above the horizon.[24]

[22]By "alternate," Marinos means "point sometimes north and sometimes south," i.e., the torrid zone falls within the belt of the globe between the two tropics (see pp. 12–13). The subsequent statement that some of the stars of Ursa Minor are not visible within the torrid zone sets the northern limit of this zone at a latitude equal to approximately 12⅖° north of the equator, that is, one degree, or 500 stades, north of Okēlis.

[23]For an observer at a terrestrial latitude $\phi$ north of the equator, a particular star will never set if the arc between it and the north celestial pole (i.e., 90° minus its declination) is less than $\phi$. The star of the constellation Ursa Minor that was farthest from the pole in Marinos' time was α UMi, the "last star of the tail," now commonly known as Polaris. Marinos cites Hipparchus as having stated that α UMi is 12⅖° from the pole, from which he concludes that the southernmost latitude at which the star never sets is 12⅖° N, and this is one degree north of his latitude for Okēlis. It is noteworthy that Marinos was not aware that the precessional motion of the stars slowly changes their declinations; and Ptolemy, who was aware of this phenomenon (cf. *Almagest* 7.3), does not draw attention to it here since it does not materially affect his criticism. In 130 B.C., the approximate date of Hipparchus' observations, α UMi was 12°28' from the pole, in good agreement with his measurement. By A.D. 100 its distance was only 11°13', so that Marinos should have set the limit of the torrid zone at the latitude of Okēlis or very slightly to the south. In our time α UMi has of course become the northernmost bright star, being less than 45' from the pole.

[24]Ptolemy does not doubt that these phenomena would actually be observed by someone south of the equator, but in this and the following sentences he denies, first, that any observations that would indicate a place lies south of the equator have ever actually been recorded, and second, that the phenomena Marinos describes are unique to places south of the equator.

In what follows he does adduce some observed phenomena, but not such as can prove his thesis in the least. Thus he says: "The people from India who sail to Limyrikē (as Diodoros of Samos says in his third [book]) see Taurus in midheaven and the Pleiades along the middle of the yard.[25] Those who set sail from Arabia to Azania direct their sail towards the south and the star Canopus [α Car], which is called *Hippos* [i.e., 'horse'] there and is in the extreme south. Stars are visible to them that we have no name for; and Sirius [α CMa] rises before Procyon [α CMi],[26] and all of Orion before the summer solstice." Now, some of these phenomena too clearly indicate habitations north of the equator (such as Taurus and the Pleiades being at zenith, since these stars are north of the [celestial] equator), while the others no more [indicate] southern [habitations] than northern. Thus Canopus can be visible to people quite far north of the Summer Tropic, and many stars that are always below the horizon for us can be above the horizon in locations south of us but still north of the equator, such as [locations] near Meroē, just as Canopus itself [can be seen] here [at Alexandria] though it is not visible to people north of us.[27] And yet the more southern people give the name *Hippos* to this [star], and not to some other [star] unknown to us. [Marinos] himself adds that by mathematical arguments it has been established that all of Orion is visible to people living on the equator before the summer solstice; and Sirius begins to rise before Procyon for those living on the equator, and as far [north] of them as Soēnē.[28] Hence neither of these phenomena is a particular characteristic of the habitations south of the equator.

[25]The first part of this sentence is puzzling, since Limyrikē is part of India, and it would be strange to speak of sailing "from India to Limyrikē." We take the phrase as an indication of the origin of the mariners, not where they began their journey. Dihle (1974, 11) sees it as a relic of a stage in earlier Greek exploration of the Indian Ocean when Limyrikē was not yet thought of as part of India.

[26]Possibly, "Canis Major rises before Canis Minor," since the stars bear the same names in Greek as the constellations that contain them. As Marinos' next example shows, the risings discussed in this passage are the *heliacal risings*, i.e., the dates in the year when stars or constellations first become visible near the eastern horizon just before dawn.

[27]Canopus does not rise above the horizon for an observer north of 37°, so that it would be visible in Egypt, but not Greece.

[28]Ptolemy's criticisms of Marinos are not intended to be obvious, but would need to be confirmed by a work such as Ptolemy's *Almagest* or his *Phaseis* (a calendar of risings and settings of a selection of bright stars). According to the *Phaseis* (Heiberg 1907, 59), Procyon is first visible in the morning at the latitude of Soēnē about July 13, just three days before Sirius; at the equator, Sirius is seen earlier. Similarly, the stars of Orion would all be visible at the equator before the summer solstice; but for the latitude of Soēnē, the *Phaseis* (Heiberg 1907, 56) gives the summer solstice itself (June 25) as the date of first appearance of one of the stars in Orion (ε Ori).

## *8. The same revision [of the latitudinal dimension], on the basis of land journeys*

Let us now turn to the journeys. For the land journey, [Marinos], counting the days of the individual marches from Leptis Magna to the country of Agisymba, reckons that [Agisymba] is 24,680 stades south of the equator. For the sea [journey], again using the days of sail from Ptolemais in Trōglodytikē to Cape Prason, he reckons [Cape Prason] also to be 27,800 stades south of the equator. In this way he moves Cape Prason and the country Agisymba—which belongs to the Aithiopians[29] and, as he himself says, does not constitute the southern limit of Aithiopia—to the frigid zone of the *antoikoumenē*; for 27,800 stades make $55\frac{3}{5}^\circ$ on the meridian, and this is as many degrees [south] from the equator as the Skythians and Sarmatians who live to the north of Lake Maiōtis are in the other direction, [and the Skythians and Sarmatians live] in this kind of [frigid] climate.

Now [Marinos] himself reduces the foregoing number of stades to less than half, that is, to 12,000 stades, which is approximately how far the Winter Tropic is from the equator. The only reasons that he gives for the reduction are the diversions from rectilinear routes and the variable speeds of the journeys, while he passes over still more fundamental and obvious [reasons] that make it clear not only that a reduction is necessary, but also [that it must be reduced] to this extent.

First, concerning the land journey from Garamē to the Aithiopians he says that Septimius Flaccus, who made a campaign out of Libyē, reached the Aithiopians after leaving the people of Garamē after marching south for three months; and Julius Maternus, who [campaigned] from Leptis Magna, leaving Garamē together with the king of the people of Garamē (who was making an expedition against the Aithiopians), after they had all marched for four months to the south, reached Agisymba, the Aithiopians' country, where the rhinoceros congregate.[31] Each of these [accounts] is implausible even on its own, both because the inland Aithiopians are not so far as a three-month journey from the people of Garamē (who are themselves more or less Aithiopians and have the same king as [the inland Aithiopians]) and because it would be absolutely pre-

[29]By "Aithiopians" Ptolemy means dark-skinned people, as becomes clear in the next chapter. See Geographical Index.

[30]The inhabitable part of the southern hemisphere corresponding to the *oikoumenē* in the northern hemisphere.

[31]Nothing definite is known about the Roman travelers Septimius Flaccus and Julius Maternus. Desanges (1978, 197–213) suggests that Flaccus led a military expedition in the reign of Domitian, c. A.D. 80, and that Maternus traveled to Agisymba c. A.D. 90 to procure rhinoceros for the imperial games at Rome. On these journeys, and the calculations Marinos and Ptolemy based on them, see Appendix A.

posterous [to imagine that] the king's expedition toward his subjects was in just one direction, from north to south, when these peoples are stretched out far on either side, to the east and west, to say nothing of making no significant pauses anywhere. For these reasons it is likely that [these] men either told travelers' tales or used the expression "to the south" for "toward the *Notos* wind" or "toward the *Lips* wind," as the locals tend to talk, misusing the rough [term] in place of the exact.

## *9. The same revision [of the latitudinal dimension], on the basis of sea journeys*[32]

Next, concerning the sail between Arōmata and Rhapta, he says that a certain Diogenes, who was one of those who sailed to India, returning the second time, was driven back when he got to Arōmata by the *Aparktias* [north] wind and had Trōglodytikē on his right for twenty-five days, and [he then] reached the lakes from which the Nile flows, slightly to the south of which is Cape Rhapta.[33] [Marinos also says] that a certain Theophilos,[34] one of those who sailed to Azania, set sail from Rhapta by the *Notos* wind, and on the twentieth day reached Arōmata. Neither of these said that it was a sail "of so many days." Theophilos said that he came to land on the twentieth day, and Diogenes [said] that he sailed along Trōglodytikē for twenty-five days; thus they reported only how many days they sailed, without reckoning how many days' sail it was,[35] taking account of the variation and shifting of the winds over such a long time. Nor [did they say] that their sail was entirely to the north or to the south. Diogenes said only that he had been driven by the *Aparktias* wind, and Theophilos, that he had set sail by the *Notos* wind, neither of them saying that the rest of the voyage maintained the same direction; for it is not to be believed that the course of the winds would be maintained for so many days.

But there is also this reason [for doubting Marinos' calculations]: whereas Diogenes traveled the interval from Arōmata to the lakes to the south of which

[32]On the use Marinos made of the voyages reviewed in this chapter, see Appendix B.

[33]Diogenes, who is otherwise unknown, must have taken the southern trade route by the southwest monsoon from Arōmata to southern India. The return voyage was begun in December or January, using the northeast monsoon to cross the open ocean and seasonal east winds to pass through the Gulf of Aden. Presumably Diogenes hit the African coast too far south to enter the gulf and was driven even further south. Ptolemy returns to the topic of the location of the lakes at the source of the Nile in 1.17.

[34]Theophilos is also otherwise unknown.

[35]Ptolemy's point is that a trip may, on the average, be so many days' sail, but sometimes might take more or fewer days. And since all the two travelers said was how many days it actually took them, not the average for all travelers, one should not treat their figures as averages. In the *Periplus Maris Erythraei* (cf. p. 27), coastal distances are often given in terms of number of days' sail.

is Cape Rhapta in twenty-five days, Theophilos sailed the interval from Rhapta to Arōmata, which is greater, in twenty days. Though Theophilos set down the travel of a day and a night as a sail of a thousand stades,[36] which [Marinos], too, follows, nevertheless [Marinos] says that the sail from Rhapta to Cape Prason, which is of many days, was set down by Dioskoros as only five thousand stades,[37] presumably because the winds on the equator are easily changeable since the shifting positions of the sun from side to side [i.e., from north to south] are more pronounced there.

Not only for these reasons should he have abandoned the number of days recorded, but also for the most manifest reason of all: the resulting computation brings the Aithiopians and the gathering place of rhinoceros into the frigid zone of the *antoikoumenē*, although all animals and plants that are on the same parallels or [parallels] equidistant from either pole ought to exist in similar combinations in accordance with the similarity of their environments.[38]

And for this very reason Marinos reduced the distance [to the southern limit of the *oikoumenē*] to just as far as the Winter Tropic. But he gave not a single logical reason for reducing by *this* amount, [even] if one were to accept (as he does) both the number of days and the [assumption] that the journeys were uniform [in direction and speed]; retaining these [data], he merely reduces the number of stades per day beyond what is reasonable or usual, until the [southern] limit reaches the parallel that he thinks it ought to reach. On the contrary, it would have been logical for one to believe that a day's travel *could* be so far, while not believing that they were uniformly constant in speed and direction. So that on the basis of these [data] we cannot obtain the distance we are looking for, but merely the fact that [this distance] is greater than the distance to the equator.[39]

Instead, [the distance could have been obtained] from some one of the more unambiguous [i.e., astronomical] phenomena—and something of this kind would

[36]Cf. 1.17 for Ptolemy's estimate of only 500 stades for a sail of a day and a night near Rhapta. Night sailing was generally to be avoided, but the *Periplus* (ch. 15; Casson 1989, 60) confirms the practice of both night and day runs along this coast.

[37]Dioskoros is otherwise unknown. It is not clear whether he, too, was a mariner or a geographical writer.

[38]I.e., if we find dark-skinned people and rhinoceros at a certain latitude south of the equator, then we should expect to find such things at a similar latitude north of the equator. The argument appears again at the end of this chapter.

[39]In other words, Marinos assumed that each day's journey was of constant length, but, rather than adopt a traditional value, he chose a constant that would produce an acceptable value for the total distance. On the other hand, Ptolemy assumes that the traditional estimate for a day's travel is valid, but only as a maximum value. Consequently the daily journeys could have varied sufficiently, in both extent and direction, to make it impossible to obtain a reasonably accurate figure for the total north-to-south distance.

also have been absolutely accurate—supposing someone making the investigation in a more mathematical manner had come across the [phenomena] characteristic to those countries. But since such research has not been made, there is nothing for it but to examine more roughly, and on the basis of a simpler [kind of evidence], what a reasonable amount for the extension [of the known world] beyond the equator would be. This is the [evidence] of the forms and colors of the local animals, from which it would follow that the parallel through the country of Agisymba, which clearly belongs to the Aithiopians, is not as far as the Winter Tropic, but lies nearer the equator. For in the correspondingly situated places on our side [of the equator], that is those on the Summer Tropic, people do not yet have the color of the Aithiopians, and there are no rhinoceros and elephants; but in places not much to the south of these, moderately black people are to be found, such as those who live in the "Thirty Schoinoi" outside Soēnē. Of the same type, too, are the people of Garamē, whom Marinos also says (and indeed for this very reason) live neither right on the Summer Tropic nor to the north, but entirely to the south of it. But in places around Meroē people are already quite black in color, and are at last pure Aithiopians, and the habitat of the elephants and more wonderful animals is there.

## *10. That one should not put the Aithiopians south of the parallel situated opposite to that through Meroē*

Thus, it would be best at this stage of the question, that is, so long as the report of those who reach there records [the presence of] Aithiopians, to draw the country of Agisymba and Cape Prason, together with whatever lies on the same parallel, approximately on the parallel correspondingly situated to the one through Meroē, that is, the parallel which is south of the equator by the same number of degrees, 16⁵⁄₁₂°, or approximately 8,200 stades. Hence the whole latitudinal breadth amounts roughly to 79⁵⁄₁₂°, or in whole numbers 80° or 40,000 stades.

We should, however, retain the interval between Leptis Magna and Garamē as Flaccus and Maternus set it, as 5,400 stades. For the twenty days belong to a return journey, which was shorter in comparison with the first, since [the return] was in a north-south direction while the [first journey] was thirty days long because of the diversions. Moreover, [Marinos] says that the travelers stated the number of stades for each day, [this number] having often been not just possible but also necessary because of the distances between the watering places. Just as one has to reserve judgment concerning great distances and those which have seldom been traveled, or not [traveled] in a way about which there is general agreement, so one should trust those which are not great but have been traveled often and by many people in a way that is agreed upon.

## *11. On the computations that Marinos improperly made for the longitudinal dimension of the* oikoumenē

The foregoing should have made it clear how far it would make sense to extend the latitudinal dimension of the *oikoumenē*. Marinos makes its longitudinal dimension bounded within two meridians that cut off fifteen hour-intervals. We think that he has also extended the eastern part of this dimension more than necessary, and that when a reasonable reduction has been applied here, too, the whole longitudinal extent does not amount quite to twelve hour-intervals, where we (like [Marinos]) set the Islands of the Blest at the westernmost limit, and the farthest parts, [namely] Sēra, Sinai, and Kattigara, at the eastern [limit].

For in the first place one should follow the numbers of stades, from place to place, set down by [Marinos] for the distance from the Islands of the Blest to the crossing of the Euphrates at Hierapolis (as if [the journey] were made along the parallel through Rhodes).[40] [This is] both because it is continually being checked and because [Marinos] has manifestly taken into account the amount by which the greater distances ought to be corrected on account of diversions and variations in the itineraries. Furthermore, [he has taken into account] the fact that one degree (of such as the great circle is 360°) contains 500 stades on the surface of the earth—in accordance with the surface measurements that are generally agreed upon—while an arc similar to [one degree of the equator] on the parallel through Rhodes (that is, the parallel 36° from the equator) contains approximately 400 stades. (We may ignore, in such a rough determination, the slight excess over [400] that follows from the [exact] ratio of the parallels.)[41]

However, we reduce according to the appropriate correction both the distance from that crossing of the Euphrates to the Stone Tower, which amounts (according to him) to 876 *schoinoi* or 26,280 stades, and that from the Stone Tower to Sēra, the metropolis of the Sēres, a journey of seven months, or [according to Marinos] 36,200 stades reckoned on the same parallel [through Rhodes].[42] For in the case of both journeys, [Marinos] has clearly not subtracted the excess resulting from diversions, and in the case of the second, he has fallen as well into the same illogicalities that he also fell into concerning the journey from the people of Garamē to Agisymba.[43] There he was compelled to subtract

[40]This distance is discussed below in 1.12; see also Appendix D.

[41]Taking cos (36°) = 0.809, one obtains for the distance 404½ stades, so that using 400 stades per degree for this parallel involves an error of about 1 percent. On the method of converting terrestrial distances in stades to degrees of longitude and latitude, see pp. 16–17.

[42]The route discussed here and in the next chapter is a version of the "Silk Road" to China. See Appendix C.

[43]By "illogicalities," Ptolemy here means Marinos' assumption that a long journey through difficult country could be made at an ideal marching speed without interruptions. In the case of the journey to Agisymba, this resulted in the initial calculation that Agisymba was 24,680 stades south of the equator, which Marinos recognized was impossible (cf. 1.8).

more than half from the number of stades added up over [a journey of] four months and fourteen days because the road journey could not have been uninterrupted over such a great time. Logically this ought also to have been the case with the seven months' journey, indeed, much more so than with the route from Garamē. After all, that journey was made by the country's king, who had (it would be reasonable to suppose) some considerable advance knowledge [of the route], and the weather was completely favorable. But the route from the Stone Tower to the Sēres is subject to bad storms (for according to Marinos' assumptions it falls on the parallels through the Hellespont and Byzantion), so that for this reason, too, there must have been numerous pauses in the journey.

Moreover, it was because of the opportunity for commerce that [the route] came to be known. Marinos says that one Maes, also known as Titianus, a Macedonian and a merchant by family profession, recorded the distance measurements, though he did not traverse it himself but sent certain [others] to the Sēres.[44] [Marinos] himself apparently did not trust merchants' reports: at least, he did not give assent to the account of Philemon, in which he has reported the longitudinal extent of the island of Hibernia [i.e., Ireland] from east to west as a twenty days' journey, because [Philemon] said that he heard it from merchants. For, [Marinos] says, these merchants do not concern themselves with finding out the truth, being occupied with their commerce; rather, they often exaggerate the distances out of boastfulness.[45] But here also the circumstance that nothing else in the seven months' journey was deemed worthy of any record or report by the travelers reveals that the length of time is a fiction.

## *12. The revision of the longitudinal dimension of the known world on the basis of journeys by land*

For these reasons, and also because the road is not on a single parallel (rather, the Stone Tower is near the parallel through Byzantion, and Sēra is south of the parallel through the Hellespont), it would appear sensible here, too, to diminish the number of stades added up from the seven months' itinerary, namely 36,200, to less than half. Let it, however, be reduced just to half, for this rough determination, so that the distance in question will be reckoned as 18,100 stades, or 45¼°. It would, after all, be absurd and unheard of, when reason dictates the same size of reduction for both the routes, to follow it in the case of the route from the people of Garamē [to Agisymba] because the refutation was staring us in the face (namely the various animals in the country of Agisymba, which could

[44]The Stone Tower would likely have been the location of a trading station where merchants from China and from western Asia exchanged their goods. Maes himself may have traveled this far.

[45]Philemon is otherwise unknown, but perhaps he was a geographical writer. Strabo (15.1.4, Loeb 7.5) expresses a similar disdain of merchants' reports.

not be moved outside their natural places), yet in the case of the route from the Stone Tower [to Sēra], not to accept the logical consequence since such a refutation did not happen to be applicable there because the environment along the whole distance would be the same, whether [the distance] be greater or smaller—just as if someone were not to act rightly, [that is], in the manner appropriate to philosophy, so long as he was not about to be caught.

The first interval, by which I mean the 876 *schoinoi* from the Euphrates to the Stone Tower, must be reduced, because of the diversions in the routes, to just 800 *schoinoi*, or 24,000 stades. For, granted that the total distance[46] of the whole [route] may be believed because it has been measured in moderately sized parts that have been much traveled, nonetheless it is obvious even from Marinos' assumptions that it has numerous detours. It is true that the route from the crossing of the Euphrates at Hierapolis through Mesopotamia to the Tigris, and from thence through the Garamaioi in Assyria and Mēdia to Ekbatana and the Caspian Gates, and to Hekatompylos in Parthia, can be situated near the parallel through Rhodes, since this parallel, according to [Marinos], too, is drawn through the countries mentioned. But the road from Hekatompylos to the city of Hyrkania must veer to the north, since the city of Hyrkania lies more or less between the parallel through Smyrna and the parallel through the Hellespont because the parallel through Smyrna is drawn right under the country of Hyrkania, while that through the Hellespont is drawn through the southern end of the Hyrkanian [Caspian] Sea, which is a little to the north of the city of the same name [i.e., Hyrkania].

Again, the road thence to Antiocheia Margianē through Areia inclines at first to the south, since Areia lies on the same parallel as the Caspian Gates, and then to the north, since Antiocheia is situated near the parallel through the Hellespont. Thence, the road to Baktra extends to the east, from there to the ascent of the range of the Kōmēdai [the road goes] to the north, and from this range to the gorge that follows upon the plains [it goes] to the south. For [Marinos] places the northern and the westernmost parts of the range, where the ascent is, on the parallel through Byzantion, and the southern and the eastern parts on the parallel through the Hellespont; this is why he says that [the route], though it leads pretty well straight east, tends to the *Notos* [south] wind. And apparently the fifty *schoinoi* from thence toward the Stone Tower incline to the north, for he says that as one ascends the gorge, the Stone Tower comes next, and from thence the mountains go off to the east and join up with the Imaon [range], which goes up from Palimbothra to the north.

Thus when the 60° that correspond to the 24,000 stades have been added to the 45¼° from the Stone Tower to the Sēres, the distance from the Euphrates to the Sēres along the parallel through Rhodes would be 105¼°. And according to

[46]Literally, "continuity" (*syneches*).

Marinos, on the basis of the individual numbers of stades that he assumes, and reckoning as on the same parallel, the distance from the meridian through the Islands of the Blest to the Sacred Cape of Spain amounts to 2½°; and that from thence to the mouth of the Baetis, and that from the Baetis to the Straits [of Hēraklēs] and Calpe each amounts again to 2½°. And, of the following [intervals], that from the Straits to Caralis in Sardinia amounts to 25°, that from Caralis to Lilybaeum in Sicily to 4½°, that from thence to Pachynus to 3°, and next that from Pachynus to Tainaros in Lakōnia to 10°, that from thence to Rhodes to 8¼°, that from Rhodes to Issos, 11¼°, that from Issos to [Hierapolis on] the Euphrates to 2½°. Thus the sum for this distance is 72°, and for the whole longitudinal extent of the known earth, from the meridian of the Islands of the Blest to the Sēres, 177¼° in total.[47]

## *13. The same revision [of the longitudinal dimension] on the basis of journeys by sea*[48]

One might also estimate that this is the size of the longitudinal dimension from the distances that [Marinos] sets down for the sail from India to the Bay of the Sinai and Kattigara, if account is taken of the [effects] of bays and variable speed of sailing, as well as of the approximate directions of the landfalls.[49] Thus [Marinos] says that after the cape marking the end of the Bay of Kolchoi, which is called Kōry, follows the Bay of Argarou, and this is 3,040 stades as far as the city of Kouroula; and the city of Kouroula is in the direction of the *Boreas* [north-northeast] wind from Kōry. Hence if a third is subtracted to account for following [the arc of] the Bay of Argarou, the crossing will amount to approximately 2,030 stades, with the irregularities of the daily sails [still incorporated in the total]. If a third is again subtracted from these [2,030 stades] to get the total distance, approximately 1,350 stades will remain in the direction of the *Boreas* wind. When this has been transferred to the [circle] parallel to the equator, and to the direction of the *Apēliōtēs* [east] wind, by subtracting half in accordance with the subtended angle, we will get the distance between the two meridians through Cape Kōry and the city of Kouroula as 675 stades. This is approximately 1⅓° since the parallels through these places do not differ significantly from the great circle [i.e., the equator].

[47]On the foregoing list of longitudinal intervals, see Appendix D.

[48]The basis of the discussion in this and the following chapter is a spice-trade route along the coasts of the Indian Ocean from India to southeast Asia (and indirectly to China). For the calculations, see Appendix E.

[49]We take *epibolē* here to mean "landfall," as it does in *Periplus* 55 (Casson 1989, 18). See pp. 16–17 for Ptolemy's procedure for converting these reported distances into longitudinal intervals in degrees.

Next, he says, the sail from the city of Kouroula is in the direction of the [sun's] winter rising [i.e., east-southeast] for 9,450 stades as far as Paloura. Again subtracting a third from this on account of the variation of the daily sails, we will obtain the total distance toward the *Euros* [east-southeast] wind, approximately 6,300 stades; and if we subtract a sixth of this in order to make the distance parallel to the equator,[50] we will find the distance between these meridians [through Kouroula and Paloura] as 5,250 stades, or 10½°.

He sets down [the width of] the Bay of Ganges as 19,000 stades [starting] from this point, and the sail across it from Paloura to Sada as 13,000 [stades] in the direction of the equinoctial sunrise [i.e., due east]. It follows that only a third of this has to be subtracted to account for the variability of the sail, so that there remain 8,670 stades, or 17⅓°, as the interval between the meridians.

Next, he makes the sail from Sada to the city of Tamala 3,500 stades in the direction of the winter sunrise. Again subtracting a third of this on account of the variation [of daily sails], we obtain 2,330 stades for a continuous course, and, because of the inclination toward the *Euros* [east-southeast] wind, we again subtract a sixth of this, to find 1,940 stades, or approximately 3⅚°, for the distance between the stated meridians.

After this, he records the traversal[51] from Tamala over the Golden Peninsula as 1,600 stades, again in the direction of the winter sunrise, so that here, too, when the same fractions have been subtracted, there remains a distance between the meridians of 900 stades, or 1⅘°. Thus the distance from Cape Kōry to the Golden Peninsula amounts to 34⅘°.

## *14. On the crossing from the Golden Peninsula to Kattigara*

Marinos does not list the numbers of stades for the sail across from the Golden Peninsula to Kattigara. However, he does say that Alexandros[52] had written that the land from thence faces the south, and those who sail along it reach the city of Zabai in twenty days; and, after sailing from Zabai in the direction of the *Notos* [south] wind and rather more to the left for a few days, there follows

[50]Deducting one-sixth implies an angle of about 34° between due east and the sun's winter rising point (equated with the direction of *Euros*). This would be astronomically correct for a much more northerly latitude (about 44°); near the equator the angle approaches its minimum, about 24°. Ptolemy probably took the direction to be simply 30° from due east regardless of one's latitude, and his deduction of one-sixth is also likely to be a rough schematic correction, adequate given the crudeness of the data.

[51]The word we translate "traversal" (*diaperama*) is rare, and occurs nowhere else in the *Geography*. It apparently means a crossing of a narrow neck of sea or land. Comparison with Ptolemy's own map suggests that at this stage the trade route took a shortcut across the Golden Peninsula.

[52]Otherwise unknown. It is not clear whether this Alexandros is reporting his own travels or those of another.

Kattigara. [Marinos] exaggerates the distance in question, taking the expression "some days" as meaning "many days," and saying that because of their multitude they could not be expressed by a number. *This*, in my opinion, is ridiculous: what number of days would be inexpressible, even if it comprised the circuit of the entire traveled world? Or what was to prevent Alexandros from saying "many" instead of "some," just as [Marinos] said Dioskoros had reported that the sail from Rhapta to Cape Prason was "many" days? One would more reasonably interpret the expression "some" as meaning "few," for we usually use this word in this way.

But in order that we should not ourselves appear to be adjusting our estimates of the distances to make them fit some predetermined amount, let us treat the sail from the Golden Peninsula to Kattigara, which comprises twenty days as far as Zabai and "some" more as far as Kattigara, in the same way as we treated the sail from Arōmata to Cape Prason, which also consists of the same number of days (twenty) to Rhapta according to Theophilos, plus "many" more to Prason according to Dioskoros. Hence we, too, in Marinos' manner, shall make the expression "some days" equivalent to "many days." Now we have shown from reasonable arguments and on the basis of the [natural] phenomena themselves that Cape Prason is on the parallel that is 16 5/12° south of the equator, while the parallel through Arōmata is 4 1/4° north of the equator,[53] so that the distance from Arōmata to Cape Prason amounts to 20 2/3°. Hence it would be reasonable for us to make the voyage from the Golden Peninsula to Zabai and from thence to Kattigara the same number [of degrees' worth of distance].

There is no need to diminish [the sail] from the Golden Peninsula to Zabai, since it is parallel to the equator because the country in between lies facing the south; but that from Zabai to Kattigara should be reduced to get the direction parallel to the equator, because the sail is toward the *Notos* wind and to the east. If we assign half the degrees to each of the intervals, because the difference between them is not clear, and again subtract a third of the 10 1/3° from Zabai to Kattigara on account of the inclination [from due east],[54] we will get the distance from the Golden Peninsula to Kattigara, as [measured] in the direction parallel to the equator, as approximately 17 1/6°. But the distance from Cape Kōry to the Golden Peninsula was shown to be 34 4/5°; hence the whole distance from Cape Kōry to Kattigara is approximately 52°.

[53]In the geographical catalogue (4.7) Ptolemy assigns Arōmata a different latitude, 6° north of the equator. The latitude given here is likely to be Marinos', retained by Ptolemy here through oversight (see Appendix B). Ptolemy's whole argument here is clearly designed to obtain by hook or by crook a longitude for Kattigara just slightly short of the preconceived figure of 180° for the breadth of the *oikoumenē*, notwithstanding what he said at the outset of this paragraph.

[54]Subtraction of one-third would correspond to an angle of about 48°, i.e., roughly 45°, between due east and the direction of sail. For the present calculation, Ptolemy evidently interprets "in the direction of the *Notos* wind and rather more to the left" as halfway between due south and due east.

But according to Marinos, the meridian through the beginning of the river Indus is a little to the west of the northernmost point of Taprobanē, which lies opposite [i.e., due south of] Cape Kōry; and the meridian through the mouths of the river Baetis is eight hour-intervals from this [meridian], or 120°, and moreover the meridian though the Islands of the Blest is 5° from this. Hence the meridian through Cape Kōry is a little more than 125° from the meridian through the Islands of the Blest, and [the meridian] through Kattigara is a little more than the sum, 177°, from that through the Islands of the Blest. This is roughly the same distance as was computed [above] along the parallel through Rhodes.

Let, however, the longitudinal dimension as far as the metropolis of the Sinai be assumed to be, in round numbers, 180° or twelve hour-intervals, since all agree that [the metropolis of the Sinai] is east of Kattigara. Thus the longitudinal dimension [along the parallel] through Rhodes amounts to approximately 72,000 stades.

## *15. On the inconsistencies in details of Marinos' exposition*[55]

We have reduced the distances in longitude to the east and in latitude to the south to this extent for the reasons that we have given. We have also concluded that the positions of the individual cities call for correction in many places where, because of the copiousness and detail of his compilations, [Marinos] gives them positions in different passages that conflict with one another or are illogical; for example, in the [features] that are believed to be "oppositely situated" [i.e., due north or south].

He says, for instance, [1] that Tarraco is opposite Caesarea Iol, although he draws the meridian through [Caesarea Iol] also through the Pyrenees, which are more than a little to the east of Tarraco. Again [2] [he says that] Pachynus is opposite Leptis Magna, and Himera opposite Thēna, yet the distance from Pachynus to Himera amounts to about 400 stades, while that from Leptis Magna to Thēna amounts to over 1,500 [stades] according to what Timosthenes[56] records. [3] Again he says that Tergestē lies opposite Ravenna, whereas Tergestē is 480 stades from the inlet of the Adriatic at the river Tileventus in the direction of the summer sunrise, and Ravenna is 1,000 stades in the direction of the winter sunrise. [4] Similarly he says that Chelidoniai lies opposite Canopus, and Akamas opposite Paphos, and Paphos opposite Sebennytos, where again he sets the stades from Chelidoniai to Akamas as 1,000, and Timosthenes sets those from Canopus to Sebennytos as 290; but this distance [between Canopus and Sebennytos] should actually have been larger if it lay between the same

[55]For explanations of the inconsistencies tersely described in this chapter, see Appendix F.

[56]Timosthenes (first half of third century B.C.) was the author of a lost work *On Harbors* that gave localities along the Mediterranean coast, distances betweeen them, and other information.

meridians [as the interval between Chelidoniai and Akamas] because it subtends a [similar] arc of a larger parallel.

Again, [5] he says that Pisae is 700 stades from Ravenna, in the direction of the *Libonotos* [south-southwest] wind, but in the division of the *klimata* and of the hour-intervals he puts Pisae in the third hour-interval and Ravenna in the second.[57] And [6] though he has said that Londinium in Britain is fifty-nine [Roman] miles north of Noviomagus, he then represents [Noviomagus] as north of [Londinium] in his division of the *klimata*. Moreover, [7] having put Athōs on the parallel through the Hellespont, he puts Amphipolis and its surroundings, which lie north of Athōs, and the mouths of the Strymōn, in the fourth *klima*, which is below the Hellespont. Similarly, [8] although almost the whole of Thrace lies below the parallel through Byzantion, he has set all Thrace's inland cities in the *klima* above this parallel. Again, [9] he says: "We shall situate Trapezous on the parallel through Byzantion"; and after showing that Satala in Armenia is sixty miles south of Trapezous, nevertheless in the description of the parallels he puts the parallel through Byzantion through Satala, and not through Trapezous.

[10] He even says that the river Nile, from where it is first seen up to Meroē, will be drawn correctly [going] from south to north. Likewise, he says that the sail from Arōmata to the lakes from which the Nile flows is effected by the *Aparktias* [north] wind. But Arōmata is quite far east of the Nile; for Ptolemais Thērōn is east of Meroē and the Nile by a march of ten or twelve days, and <the Bay of Adoulis is... stades> from Ptolemais, and the straits between the peninsula of Okēlis and Dērē are 3,500 stades from Ptolemais, and the cape of Great Arōmata is 5,000 stades to the east of these.[58]

## *16. That certain matters escaped [Marinos'] notice in the boundaries of the provinces*

Some things have escaped his notice also in the definition of the boundaries: for example [11] when he has the whole of Mysia bordered by the Sea of Pontos to the east, but he has Thrace border Upper Mysia to the west; and [12] when he has Italia bordered to the north by Rhaetia and Noricum but also by Pannonia, whereas [he has] Pannonia [border] only Dalmatia, and no longer Italia, to the south. [He describes] [13] the inland Sogdians and Sakai as being neighbored by India to the south. However, he does not draw through these peoples the two parallels (namely, the parallels through the Hellespont and through Byzantion) that are [immediately] north of the Imaon range—i.e., the most northerly part

[57]See Textual Notes (Appendix G).

[58]The words between angle brackets have been restored on geographical grounds; see Appendices F and G (Textual Notes).

of India. On the contrary, the first [parallel that he draws through these peoples] is the one through the middle of Pontos.

## *17. On the inconsistencies between [Marinos] and the reports of our time*

Marinos did not notice these and similar things, either because his compilations were so voluminous and treated [various] topics separately, or because he did not have time in his final publication, as he himself says, to draw a map, which is the only way that he could have corrected the *klimata* and the hour-intervals. In some matters he is also not in agreement with present-day accounts. For example, he places the Bay of Sachalitēs to the west of Cape Syagros, when absolutely everyone who has sailed through these places agrees with our opinion that the country of Sachalitēs in Arabia and the bay of the same name are east of Syagros.

Again, he puts "Simylla" (the trading post in India) west not only of Cape Komaria, but also of the river Indus. But there is a consensus among those who have sailed there and visited the places over a long period, as well as among those who have come to us from there, that [this place] is just south [and not west] of the mouths of the river, and it is called "Timoula" by the natives.

From these people we have also learned other details about India, especially about the provinces and the more remote parts of this country as far as the Golden Peninsula and from that point on to Kattigara. First, they agree in reporting that the sail is eastward when one is sailing there, and westward when departing.[59] [Second], they agree that the direction varies and the journeys are unequal in duration. [Third], the country of the Sēres and the metropolis of the Sēres lie above [i.e., to the north of] the Sinai. To the east of these is an unknown country that has reedy lakes in which reeds grow so densely that one is borne by [the reeds] as one crosses [the lakes]. They further [agree] that not only is there a route from [the Sēres] to Baktria via the Stone Tower, but also to India via Palimbothra;[60] and the route from the metropolis of the Sinai to the station at Kattigara is to the west and south. Consequently [this route] does not fall along the meridian through the Sēres and Kattigara, as Marinos says, but on [meridians] that are east of it.

And we learn from the merchants who have crossed from Arabia Felix to Arōmata and Azania and Rhapta (they give all these [places] the special name

[59]It is not clear whether Ptolemy is talking about the last stage of sail to Kattigara, or the general trend of the entire route. On his map, Ptolemy accepts Marinos' contention that the crossing of the Great Bay from Zabai to Kattigara is slightly east of south.

[60]The trade routes branching off the Silk Road and into India were already known to the author of the *Periplus* (64; Casson 1989, 91).

Barbaria) that the sail is not exactly to the south; but rather this part is to the west and south, while they make the sail across from Rhapta to Prason toward the east and south. And the lakes from which the Nile flows are not right by the sea but quite far inland.

[We learn] also that the sequence of beaches and bluffs to Cape Prason from the Cape of Arōmata is different from what it is according to Marinos, and the sail of a day and night there does not amount to many stades, because of the swift changeability of the winds at the equator, but is generally four or five hundred stades. [We learn also that] immediately following Arōmata is a first bay, and in it, after a day's travel from Arōmata, is the town of Panōn and the trading post Opōnē, which is six days' journey from the town. Another bay, which is the beginning of Azania, follows after this trading post, and at its beginning is situated the headland of Zingis and the mountain Phalangis, which has eight peaks. This bay alone is called "Bluff"; it takes two days and nights to cross. Following this is the Little Beach, whose crossing requires three intervals [i.e., day's or night's sails], and then the Great Beach, which requires five intervals to cross. To cross the two together requires four days and nights in all. Another bay is adjacent to these, in which, after a sail of two days and nights, there is the trading post called Essina. Then comes the anchorage of Sarapiōn after one day's sail, and then begins the bay leading to Rhapta, with a crossing time of three days and nights. At its beginning is a trading post called Toniki, and by Cape Rhapton is the river called Rhaptos, and a metropolis of the same name [Rhapta], which is a little distance from the sea. The bay from Rhapta to Cape Prason is very big and not deep, and barbarous cannibals live about it.

## *18. On the inconvenience of Marinos' compilations for drawing a map of the* oikoumenē

We shall make this the end of our outline of the things that need some attention in the research [of Marinos] itself, lest it should seem to anyone that we are undertaking a prosecution rather than a revision; for everything will be clear to us in the guide to the individual parts [of the map]. We must still investigate the method of drawing the map. This undertaking can take two forms: the first sets out the *oikoumenē* on a part of a spherical surface, and the second on a plane. The object in both is the same, namely convenience; that is, to show how, without having a model already at hand, but merely by having the texts beside us, we can most conveniently make the map. After all, continually transferring [a map] from earlier exemplars to subsequent ones tends to bring about grave distortions in the transcriptions through gradual changes.

If this method based on a text did not suffice to show how to set [the map] out, then it would be impossible for people without access to the picture to accomplish their object properly. And in fact this is what happens to most people [who try to draw] a map based on Marinos, since they do not possess a model based on his final compilation; instead they draw on his writings and err in most respects from the consensus of opinion, because his guide is so hard to use and so poorly arranged, as anyone who tries it can see.

For example, one has to have the position in longitude and latitude for each locality that is to be marked, if one is to place it where it belongs. But this cannot be found directly in [Marinos'] compilations: rather, [one finds] separately, in one place maybe just the latitudes, say in [the section on] the setting out of the parallels, and in some other place just the longitudes, say in the [section on] inscribing the meridians. What is more, the same [localities] are not found in each section: the parallels are drawn through some places and the meridians through others, so that such localities lack one or the other position [i.e., longitude or latitude]. In general, one needs to have practically all [Marinos'] writings to make the investigation for each [locality] that is to be set down, because something different is said about the same [locality] in every one of them. If we did not check each category [of information] that is listed for each [locality], we would inadvertently err in many matters that ought to be checked. Again, when one is putting the cities in their positions, one might have an easier time labeling those that are on the coast, since in general some [indication of] position is noted for them, but this is not so for the inland ones, since their relative positions with respect to each other or with respect to [the cities on the coast] are not indicated, with few exceptions—and in *these* instances sometimes [only] the longitude is defined, sometimes [only] the latitude.

## *19. On the convenience of our catalogue for making a map*

We have thus taken on a twofold task: first to preserve [Marinos'] opinions [as expressed] through the whole of his compilation, except for those things that need some correction; second to see to it that the things that he did not make clear will be inscribed as they should be, so far as is possible, using the researches of those who have visited the places, or their positions [as recorded] in the more accurate maps. We have taken care also that the method should be convenient. Hence we have written down for all the provinces the details of their boundaries (i.e., their positions in longitude and latitude), the relative situations of the more important peoples in them, and the accurate locations of the more noteworthy cities, rivers, bays, mountains, and other things that ought to be in a map of the *oikoumenē*. [By "locations" I mean] the number of degrees (of such as the great circle is 360) in longitude along the equator between the meridian drawn through the place and the meridian that marks off the western limit [of the *oikoumenē*], and the number of degrees in latitude between the parallel drawn through the place and the equator [measured] along the meridian. In this way we will be able to establish the position of each, and through accuracy in particulars [we will be able to establish] the positions of the provinces themselves with respect to each other and to the whole *oikoumenē*.

## *20. On the disproportional nature of Marinos' geographical map*

Each of the two approaches [to map-making] is characterized in the following way. Making the map on the globe gets directly the likeness of the earth's shape, and it does not call for any additional device to achieve this effect; but it does not conveniently allow for a size [of map] capable of containing most of the things that have to be inscribed on it, nor can it permit the sight to fix on [the map] in a way that grasps the whole shape all at once, but one or the other, that is, either the eye or the globe, has to be moved to give a progressive view [of the whole]. Drawing the map on a plane eliminates these [difficulties] completely; but it does require some method to achieve a resemblance to a picture of a globe, so that on the flattened surface, too, the intervals established on it will be in as good proportion as possible to the true [intervals].

Marinos paid considerable attention to this problem, and found fault with absolutely all the [existing] methods of making plane maps. Nonetheless, he himself turns out to have used the one that made the distances least proportionate. He made the lines that represent the parallel and meridian circles all straight lines, and also made the lines for the meridians parallel to one another, just as most [mapmakers] have done; but he kept only the parallel through Rhodes proportionate to the meridian in accordance with the approximate ratio of 5:4 that applies to corresponding arcs on the sphere (that is, the ratio of the great circle to the parallel that is 36° from the equator), giving no further thought to the other [parallels], neither for proper proportionality nor for a spherical appearance. Now when the line of sight is initially directed at the middle of the northern quadrant of the sphere, in which most of the *oikoumenē* is mapped, the meridians can give an illusion of straight lines when, by revolving [the globe or the eye] from side to side, each [meridian] stands directly opposite [the eye] and its plane falls through the apex of the sight.[61] The parallels do not do so, however, because of the oblique position of the north pole [with respect to the viewer]; rather, they clearly give an appearance of circular segments bulging to the south.

In the next place, although in both truth and appearance the same meridians cut off similar but unequal arcs on the parallels of different sizes, and always greater [arcs] on those nearer the equator, [Marinos] makes them all equal, stretching the intervals in the *klimata* north of the parallel through

[61]I.e., the vertex of the cone of visual rays; see p. 57 n. 2.

Rhodes more than they are in truth, and making the southern ones smaller. Hence [the arcs] no longer match the numbers of stades that he has set down, but those on the equator fall short by about one-fifth (which is the amount by which the parallel through Rhodes falls short of the equator [in length]), and those on the parallel through Thulē are in excess by four-fifths (the amount by which the parallel through Rhodes exceeds that through Thulē—for in such units as the equator is 115, the parallel 36° from the equator and drawn through Rhodes is 93, and the parallel 63° [from the equator] and drawn through Thulē is 52).[62]

## *21. On the things that should be preserved in a planar map*

For these reasons it would be well to keep the lines representing the meridians straight, but [to have] those that represent the parallels as circular segments described about one and the same center, from which (imagined as the north pole) one will have to draw the meridian lines. In this way, above all, a semblance of the spherical surface will be retained in both its actual disposition and its visual effect, with the meridians still remaining untilted with respect to the parallels and still intersecting at that common pole. Since it is impossible to preserve for all the parallels their proportionality on the sphere, it would be adequate [1] to keep this [proportionality] for the parallel through Thulē and the equator (so that the sides that enclose our [*oikoumenē*'s] latitudinal dimension will be in proper proportion to their true magnitudes), and [2] to divide the parallel that is to be drawn through Rhodes (on which most of the investigations of the longitudinal distances have been made) in proportion to the meridian, as Marinos does, that is in the approximate ratio of similar arcs of 5:4 (so that the more familiar longitudinal dimension of the *oikoumenē* is in proper proportion to the latitudinal dimension). We will make clear how these things will be done, after first setting out how a map should be made on the globe.

## *22. On how one should make a map of the* oikoumenē *on a globe*

The size of the [globe] should be determined by the number of things that the map-maker intends to inscribe [on it]; and this depends on his competence and ambition, since the larger the globe is, the more detailed and at the same time the more reliable [the map] will prove to be. Whatever size it may be, we are to take its poles and accurately attach through them a semicircle very slightly separated from the [globe's] surface, so that it only just avoids rubbing against it when it is turned. Let the semicircle be narrow in order not to obstruct many

[62]See 1.24, p. 86 n. 68, for the derivation of the units on which these approximate ratios are based.

localities; and let one of its edges pass precisely through the points [representing] the poles, so that we can use it to draw the meridians. We divide [this edge] into 180 parts and label them with the [corresponding] numbers, starting from the middle division, which is going to be at the equator. Similarly, we draw the equator and divide one of its semicircles into the same number, 180, of divisions, and inscribe [their] numbers on this [semicircle] too, starting from the endpoint through which we are going to draw the most western meridian.

Now we will make the map, on the basis of the degrees of longitude and latitude recorded in [the present] writings for each marked locality, using the divisions of the semicircles of the equator and the moving meridian. We move this along to the indicated degree of longitude, that is, to the division of the equator containing the number, and take the interval in latitude from the equator directly from the divisions of the [moving] meridian; and we make a mark at the indicated number, just as we did in inscribing the stars on the solid [celestial] globe.[63]

In the same way it will be possible to draw the meridians at as many degrees of longitude as we choose, using the divided edge of the ring directly as a ruler, and [to draw] the parallels at as many intervals [from the equator] as will produce a suitable spacing, by fixing the instrument that is to draw them next to the number on the [divided] edge [of the meridian] that indicates the appropriate interval and revolving it with the [meridian] ring itself as far as the meridians that mark the limits of the known world.

## *23. List of the meridians and parallels to be included in the map*

These [limiting meridians] will enclose twelve hour-intervals according to what has been demonstrated [above].[64]

However, we have decided it is appropriate [to the size of the map] to draw the meridians at intervals of a third of an equinoctial hour, that is, at intervals of five of the chosen units [i.e., degrees] of the equator, and [to draw] the parallels north of the equator as follows:[65]

[1.] The first parallel differing [in length of longest daylight] from [the equator's twelve hours] by ¼ hour, and distant [from the equator] by

[63]This refers to *Almagest* 8.3, where Ptolemy describes the construction of a globe representing the constellations on the celestial sphere.

[64]In 1.11–14. Note that this sentence is really part of the closing paragraph of the preceding chapter. We have transposed the sentence that follows this in all manuscripts to its proper place at the end of the chapter.

[65]The following results are extracted (with latitudes rounded to the nearest twelfth of 1°) from the list of astronomical characteristics of the significant parallels in *Almagest* 2.6. The list incorporates Ptolemy's seven *klimata*.

4¼°, as established approximately by geometrical demonstrations.

[2.] The second, differing by ½ hour, and distant 8 5/12°.

[3.] The third, differing by ¾ hour, and distant 12½°.

[4.] The fourth, differing by 1 hour, and distant 16 5/12°, and drawn through Meroē.

[5.] The fifth, differing by 1¼ hours, and distant 20¼°.

[6.] The sixth, which is on the Summer Tropic, differing by 1½ hours, and distant 23 5/6°, and drawn through Soēnē.

[7.] The seventh, differing by 1¾ hours, and distant 27 1/6°.

[8.] The eighth, differing by 2 hours, and distant 30⅓°.

[9.] The ninth, differing by 2¼ hours, and distant 33⅓°.

[10.] The tenth, differing by 2½ hours, and distant 36°, and drawn through Rhodes.

[11.] The eleventh, differing by 2¾ hours, and distant 38 7/12°

[12.] The twelfth, differing by 3 hours, and distant 40 11/12°.

[13.] The thirteenth, differing by 3¼ hours, and distant 43 1/12°.

[14.] The fourteenth, differing by 3½ hours, and distant 45°.

[15.] The fifteenth, differing by 4 hours, and distant 48½°.

[16.] The sixteenth, differing by 4½ hours, and distant 51½°.

[17.] The seventeenth, differing by 5 hours, and distant 54°.

[18.] The eighteenth, differing by 5½ hours, and distant 56°.

[19.] The nineteenth, differing by 6 hours, and distant 58°.

[20.] The twentieth, differing by 7 hours, and distant 61°.

[21.] The twenty-first, differing by 8 hours, and distant 63°, which is drawn through Thulē.

[22.] And another parallel will be drawn south of the equator, containing a difference of ½ hour, and it will pass through Cape Rhapton and Kattigara, and be approximately the same number of degrees, 8 5/12°, from the equator as the oppositely situated [localities north of the equator with the same longest daylight].

[23.] And the parallel that marks the southern limit will also be drawn; it is as far south of the equator as the parallel through Meroē is north of it.[66]

## *24. Method of making a map of the* oikoumenē *in the plane in proper proportionality with its configuration on the globe*

For the map on the [planar] surface, our procedure for [maintaining] the proper proportionality of the extreme parallels will be as follows [Fig. 14].

[66]This sentence follows the first sentence of this chapter in the manuscripts, but clearly belongs here.

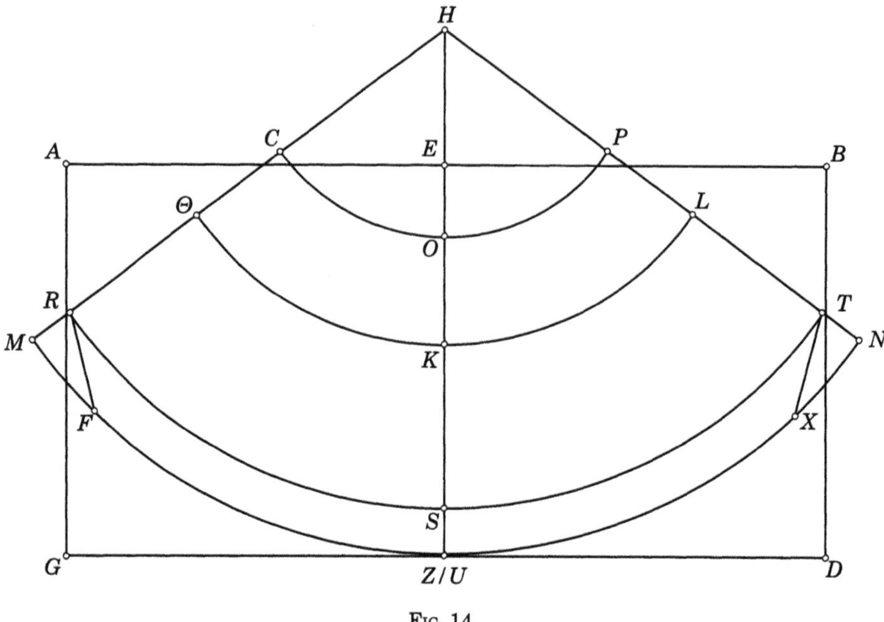


FIG. 14.

*[Ptolemy's first projection]*

*[Stage 1: Preparation of the rectangular surface on which the map is to be drawn; construction of the central meridian, the common intersection of all meridians, and the parallel through Rhodes.]* Let us fashion a [planar] surface in the shape of a rectangular parallelogram *ABGD*, with side *AB* approximately twice *AG*. Let line *AB* be assumed to be in the top position; this is going to be at the north end of the map. Then we will bisect *AB* by the perpendicular straight line *EZ*, and attach a rule *EH* to [*AB*], of suitable size and perpendicular to [*AB*] so that the line [*EH*] down the middle of its length is in a straight line with *EZ*.[67] Let there be taken on it [a length] *EH* of 34 [units] such that straight line *HZ* is 131⅚,[68] and with center *H* and radius [to reach] the point 79 units away on *HZ*, we will describe a circle *ΘKL*, which will represent the parallel through Rhodes.

[67]This rule is a temporary attachment to the surface on which the map will be drawn, to provide a base for point *H*, which will be the intersection of all the meridians and center of all the parallels of latitude in the map.

[68]These units will be equivalent to degrees of latitude on the lines representing the meridians and degrees of longitude along the circle representing the equator. The unit is chosen so that the radius of the arc representing the equator (115 units) exceeds that of the arc representing the northernmost parallel through Thulē (52 units) by 63 units, corresponding to the 63° of latitude between the equator and the parallel through Thulē. *HZ* is made equal to 115 units plus 16⅚ for the latitude of the southernmost parallel bounding the *oikoumenē*, so that this parallel will just graze the bottom of the rectangle. The length of 34 units for *EH* seems to have been empirically chosen to accommodate the largest map in the given rectangle without excessive truncation of the corners at *C* and *P*.

[*Stage 2: Construction of meridian lines at five-degree intervals.*] For the limits of the longitude, which comprise six hour-intervals on each side of *K*, we take an interval of 4 units on the middle meridian *HZ* (i.e., five degrees on the parallel through Rhodes because of the approximate ratio of 5:4 between the great circle and [the parallel]), and we count off eighteen intervals of this size on each side of *K* along arc *ΘKL*. [In this way] we get the points through which the meridians that will enclose the intervals of one-third of an hour will have to be drawn from *H*, and consequently also the [meridians] marking off the limits [of longitude], namely *HΘM* and *HLN*.

[*Stage 3: Construction of arcs representing the equator and limiting parallels, and the other parallels.*] Next the parallel *COP* through Thulē will be drawn, with radius 52 units from *H* on *HZ*, and the equator *RST*, [with radius] 115 units from *H*, and the parallel *MUN*[69] oppositely situated to the parallel through Meroē—this is the farthest south [of the parallels], [with radius] 131⁵⁄₁₂ units from *H*. Hence the ratio of *RST* to *COP* will amount to 115:52, in agreement with the ratio of these parallels on the globe, since *HO* is 52 of such units as *HS* was assumed to be 115, and as *HS* is to *HO*, so is arc *RST* to [arc] *COP*.

Also, the interval *OK* of the meridian, that is, the [interval] from the parallel through Thulē to that through Rhodes, will turn out to be 27 units; and *KS*, the [interval] from the parallel through Rhodes to the equator, will turn out to be 36 of the same [units]; and *SU*, the [interval] from the equator to the parallel oppositely situated to that through Meroē, will turn out to be 16⁵⁄₁₂ of the same [units].[70] Moreover, of such [units] as the latitudinal dimension *OU* of the known world is 79⁵⁄₁₂ (or as a round number, 80), *ΘKL*, the middle interval in longitude [measured along the parallel through Rhodes], will be 144, in agreement with the hypotheses derived from the demonstrations [in 1.7–14]—i.e., the 40,000 stades of latitude have approximately the same ratio to the 72,000 stades of longitude on the parallel through Rhodes as [the ratio 79⁵⁄₁₂:144].

We will also draw the rest of the parallels, if we choose, again using *H* as center and radii [extending to the points] as many units away from *S* as the [numbers] set out in the [list of] distances from the equator [1.23].

[*Stage 4: Inflection of meridian lines south of the equator.*] Instead of having the lines representing the meridians straight as far as parallel *MUN*, we can have them [straight] just as far as the equator *RST*; then, dividing arc *MUN*

[69]If one has followed Ptolemy's instructions, the midpoint of this arc will be point *Z*. Nevertheless, this point is consistently referred to as *U*, as if it were distinct from *Z*. The violation of the alphabetic order in which Ptolemy assigns letters to the points in his figure suggests that he originally called this point *Z*, and then changed his mind. He may have realized that one might choose to draw the map on a larger surface than he specifies, with room for margins; then, point *Z* at the bottom of the rectangle would not be the same as point *U* at the bottom of the map.

[70]In this paragraph Ptolemy wishes to bring home to the reader the extent to which he has succeeded in attaining the preservation of proportions of distances that is his goal.

into parts that are equal and equal in number to [the parts] established on the parallel through Meroē,[71] [we can] draw between these divisions and the [divisions] on the equator the straight lines for the meridians that fall between [the parallel and the equator] (e.g., lines *RF* and *TX*), so that the bending away on the other, south side of the equator is in some way apparent from the inflection [of the meridian lines] incorporated [in the map].

[*Stage 5: Drawing of the map.*] Next, to make the labeling of the localities that are to be included convenient, we will also make a narrow little ruler, equal in length to *HZ* (or just to *HS*),[72] and peg it to *H* so that as it is revolved along the whole longitudinal dimension of the map, one of its edges will exactly fit the straight lines of the meridians because it is cut away so as to be in line with the middle of the pole. We divide this edge into the 131 5/12 parts corresponding to *HZ* (or the 115 parts corresponding to just *HS*), and label the numbers starting from the division at the equator in order not to divide the middle meridian of the map into all the [115] parts and label them [all], thereby making a mess of the inscriptions of the localities that are to be made next to [the middle meridian]. It will also be possible to draw the parallels using these [marks].[73]

Then we divide the equator, too, into the 180° of the twelve hour-intervals, and annex the numbers starting at the westernmost meridian. And we shift the edge of the ruler in each case to the indicated degree of longitude, and, using the divisions on the ruler, we arrive at the indicated position in latitude as required in each instance, and make a mark just as we have explained for the globe.

[*Ptolemy's second projection*]

We could make the map of the *oikoumenē* on the [planar] surface still more similar and similarly proportioned [to the globe] if we took the meridian lines, too, in the likeness of the meridian lines on the globe, on the hypothesis that the globe is so placed that the axis of the visual rays passes through both [1] the intersection nearer the eye of the meridian that bisects the longitudinal dimension of the known world and the parallel that bisects its latitudinal dimension, and also [2] the globe's center.[74] In this way the oppositely situated limits [of

[71]That is, the whole length of arc *MUN* is not used, but only a part of it equal in length to the arc representing the parallel through Meroē.

[72]If Ptolemy's suggestion for inflecting the meridians at the equator is followed, this ruler will be useless for localities south of the equator.

[73]This would be done, as with the globe, by attaching the pen to the appropriate mark on the moving ruler.

[74]Since (as Ptolemy points out presently) the parallel through Soēnē (i.e., the Summer Tropic) is close to halfway between the northernmost and southernmost limits of the *oikoumenē*, Ptolemy means that the eye should be imagined as directly above a point that lies both on this parallel and on the central meridian of the map. Ptolemy specifies the intersection "nearer the eye" to avoid ambiguity, since any parallel and meridian intersect twice on opposite sides of the globe. For the

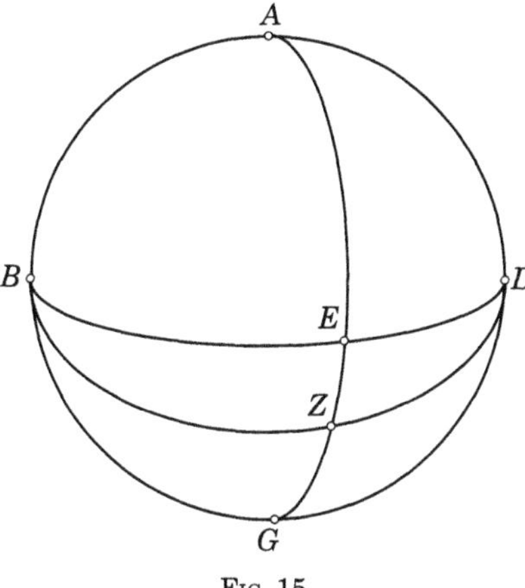


FIG. 15.

latitude and longitude] will be taken in and perceived by the visual rays at equal distances.

[*Stage 1: Determination of an appropriate point to serve as the common center of the arcs representing the parallels.*[75]] First, [we want] to establish the magnitude of inclination of the parallel circles with respect to the plane that is perpendicular to the meridian in the middle of the longitude and [that passes] through both the stated intersection [of the bisecting meridian and parallel] and the sphere's center. Let us imagine [Fig. 15] the great circle *ABGD* that delimits the visible hemisphere; the semicircle *AEG* of the meridian that bisects the hemisphere; and point *E*, which is the intersection nearer the eye of the [central meridian] and the parallel bisecting the latitudinal dimension. And let there be described through *E* another semicircle of a great circle, *BED*, perpendicular to *AEG*. Obviously the plane of [*BED*] lies along the axis of the visual rays. Let arc *EZ* be measured off as 23⅚° (since the equator is this many [degrees] from the parallel through Soēnē, which is approximately the middle of the latitudinal breadth), and let the semicircle *BZD* of the equator be de-

---

theory of visual rays, see p. 57 n. 2. The axis of the rays is the line in the middle of the field of vision that points to an object when we look straight at it; this concept is not present in Euclid's *Optics* but plays a prominent role in Ptolemy's *Optics*.

[75]In the following, Figure 15 provides a schematic picture of the globe (*not* viewed from the eyepoint above the intersection of the central meridian and parallel, which is point *E* in the figure). Ptolemy wishes to point out that there is a great circle through *E* and perpendicular to the central meridian *AEG* that a viewer directly above *E* will see as a straight line at right angles to the central meridian, which will also appear as a straight line. He assumes that the viewer is at a great distance, so that the part of the globe that the eye can see is a complete hemisphere. Because of this, the equator and the great circle through *E* will intersect at the edges of the visible part of the globe, and they will be seen respectively as an arc bulging toward the south and the chord subtending the arc. These appearances will be retained in the map.

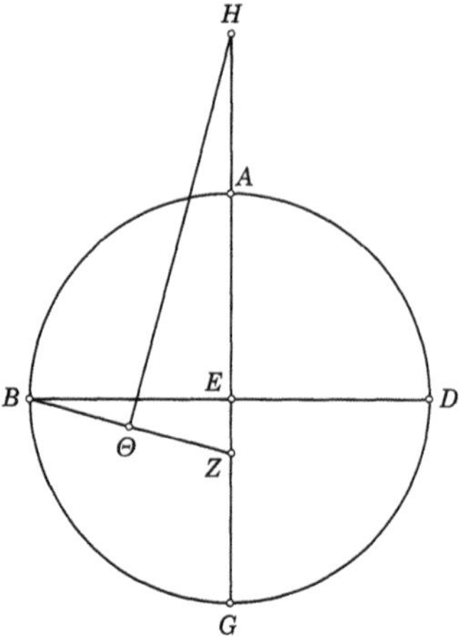


FIG. 16.

scribed through $Z$. Then the plane of the equator and the [planes] of the other parallels will appear inclined with respect to the [aforesaid plane] through the axis of the visual rays at [the angle of] arc $EZ$, which is $23\frac{5}{6}°$.

Now [Fig. 16] let $AEZG$ and $BED$ be imagined as straight lines representing arcs, such that $BE$ has a ratio to $EZ$ of $90{:}23\frac{5}{6}$.[76] And let $GA$ be produced, and let the center about which the circular segment $BZD$ is to be described be at $H$, and let it be required to find the ratio of $HZ$ to $EB$.

Let straight line $ZB$ be drawn, and bisected at $\Theta$, and let $\Theta H$ (which is of course perpendicular to $BZ$)[77] be drawn. Then since $EZ$ was assumed to be $23\frac{5}{6}$ of such [units] as straight line $BE$ is 90, the hypotenuse $BZ$ will be $93\frac{1}{10}$ of the same [units]. And angle $BZE$ will be $150\frac{1}{3}$ of such [units, i.e., half-degrees] as two right angles are 360, and the remaining angle $\Theta HZ$ will be $29\frac{2}{3}$ of the same [half-degrees].[78] Consequently the ratio of $HZ$ to $Z\Theta$ is $181\frac{5}{6}$: $46\frac{11}{20}$.[79] But of

[76]Figure 16 is the skeleton of the map, with points corresponding in meaning to the similarly named points in Figure 15. Since distances along the central meridian are to be drawn in correct proportion to the corresponding intervals on the globe, and $BED$ represents half a great circle (i.e., the equivalent of 180° of the central meridian), Ptolemy takes $\frac{1}{90}$ of $BE$ as equal to 1° along the meridian, and places $Z$ (representing the central point of the equator) as many of these "degrees" south of $E$ as the equator is south of Soēnē. The equator is going to be drawn as a circular arc passing through $B$, $Z$, and $D$, and the center $H$ of this arc is determined as the intersection of the perpendicular bisectors of the two chords $BZ$ and $BD$. By hypothesis, $H$ will be the center of all the arcs representing the parallels on the map.

[77]$\Theta H$ is perpendicular to $BZ$ because $BZ$ is a chord of circle $BZD$.

[78]This is an exercise in trigonometry similar to many in the *Almagest*. A circle is tacitly imagined as circumscribing triangle $BZE$; obviously $BZ$ is a diameter of this circle, and angle $BZE$ is half the angle subtended at the circle's center by chord $EB$ (or equivalently, as many half-degrees as the angle at the center is in degrees). Ptolemy used his table of chords (*Almagest* 1.11) to get this angle.

[79]$Z\Theta$ is half $BZ$, which has been found to be $93\frac{1}{10}$ units. Moreover $HZ{:}Z\Theta = BZ{:}ZE$ (by similar triangles), so that $HZ$ is approximately $181\frac{5}{6}$ units.

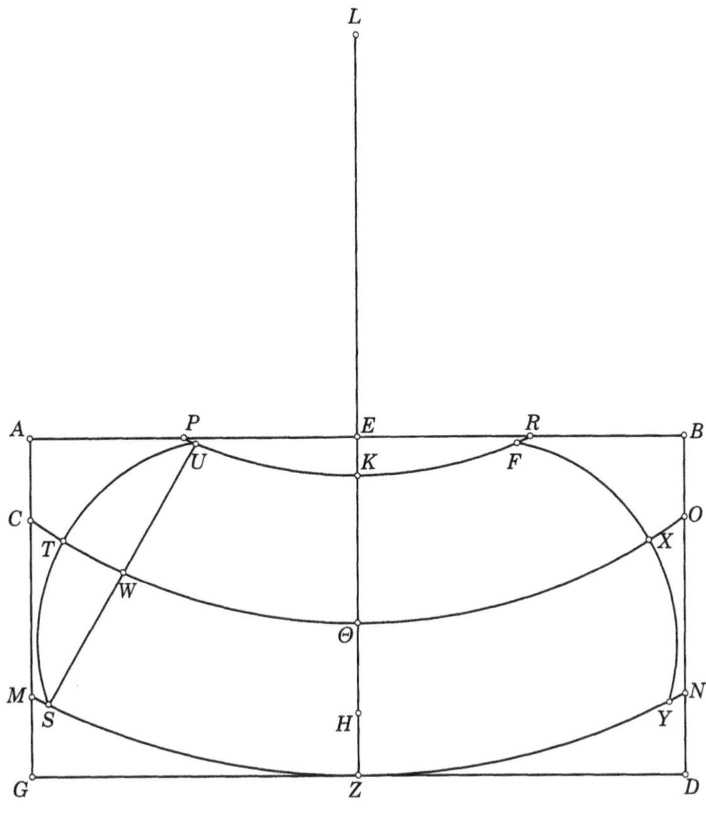


FIG. 17.

such [units] as $\Theta Z$ is $46\frac{11}{20}$, straight line *BE* is 90; so that also of such [units] as straight line *BE* is 90 (and *ZE* is $23\frac{5}{6}$ of the same), we will have straight line *HZ* too as $181\frac{5}{6}$. And [we will thus obtain] point *H*, about which all the parallels in the plane map are to be described.

[*Stage 2: Construction of the arcs for the parallels.*] Now that these things have been established, let the [plane] surface *ABGD* be set out [Fig. 17] with *AB* again being twice *AG*, and *AE* equal to *EB*, and *EZ* at right angles to [*AEB*].[80] Also let some straight line equal to *EZ* be divided into the 90 units [corresponding to the degrees] of the quadrant. Let *ZH* be taken with length $16\frac{5}{12}$ units, and $H\Theta$ with length $23\frac{5}{6}$ units, and *HK* with length 63 units. If *H* is assumed to be on the equator, $\Theta$ will be the point through which the parallel through Soēnē (which is approximately in the middle of the latitudinal dimension) will be drawn; and *Z* will be the point through which the parallel will be drawn that marks the southern limit and is opposite to the parallel through Meroē, and *K* will be the point through which the parallel will be drawn that marks the northern limit and passes through the island of Thulē.

[80]Points *L*, $\Theta$, and *H* in Figure 17 correspond respectively to *H*, *E*, and *Z* in Figures 15 and 16.

We now produce [line *EZ*'s] extension *HL* with length 181⅚ of the same units (or for that matter just 180 units, since the map will not be significantly different on this account). And with center *L* and radii [extending to] *Z*, *Θ* and *K*, we describe arcs *PKR*, *CΘO*, and *MZN*. The proper pattern of inclination of the parallels with respect to the plane through the axis of the visual rays will thus have been preserved, since here, too, [as in the hypothetical view of the globe], the axis [of the visual rays] ought to point to *Θ* and be at right angles to the plane of the map, so that the oppositely situated limits of the map will again perceived by the sight as equidistant [from the center].

[*Stage 3: Construction of the arcs for the meridians.*] The longitudinal dimension should be proportional to the latitudinal dimension. On the globe, of such [units] as the great circle is 5, the parallel through Thulē amounts to approximately 2¼, and that through Soēnē 4 7/12, and that through Meroē 4⅚. One has to place eighteen meridians at intervals of one-third hour on each side of the meridian line *ZK* to complete the semicircles [of the parallels of latitude] contained by the total longitudinal dimension, so that we will take, on each of the three parallels that have been set out, segments equivalent to 5°, i.e., one-third of an hour-interval. [Thus] we will make the divisions from *K* at intervals of 2¼ units such as straight line *EZ* is 90, and from *Θ* at intervals of 4 7/12, and from *Z* at intervals of 4⅚. Then we will draw the arcs to represent the remaining meridians through [each set of] three corresponding points, e.g., the [meridians] that are to mark the limits of the whole longitudinal dimension, *STU* and *FXY*.[81] We shall then add the rest of the [arcs] representing the remaining parallels, with center again *L* and radii [extending to] the divisions on *ZK* according to their distances from the equator.

It is immediately obvious how such a map is more like the shape on the globe than the former map. For there [i.e., on the globe], too, when the globe is stationary and not turned about (which is necessarily the case with [the situation represented on] the [planar] surface), since the sight is directed toward the middle of the map, a single meridian, [namely] the one in the middle, would be in the plane through the axis of the visual rays and so would give the illusion of a straight line; whereas the [meridians] on either side of it all appear curved with their concavities toward it, and the more so the farther from it they are. Here, too, [in the present map] one will retain this [appearance] with the proper relative curvatures. Moreover, the proportionality of the parallel arcs with respect to each other preserves the proper ratio, not just for the equator and the parallel through Thulē (as in the former [map]), but also as very nearly as possible for the other [parallels], as anyone can discover who makes the experiment.

[81]Ptolemy does not explain how these arcs are to be drawn. For the meridians close to the central meridian, the curvature is so slight that it would be impracticable to use a compass.

And [the ratio] of the total latitudinal dimension to the total longitudinal dimension [will be preserved] again not only for the parallel drawn through Rhodes (as in the former [map]), but [at least] roughly for absolutely all [the parallels]. For if here, too, as in the former drawing, we draw the straight line *SWU*, arc *ΘW* will obviously make a smaller ratio to [arcs] *ZS* and *KU* than the correct ratio in this map, which was obtained using all of [arc] *ΘT* (imagined as being along the equator). And if we make this [arc *ΘW*] in correct ratio to the latitudinal interval *ZK*, then *ZS* and *KU* will be greater than the [arcs] that are in correct ratio to *ZK* [on these parallels], just as *ΘT* is. Or if we keep *ZS* and *KU* in correct ratio to *ZK*, then *ΘW* will be less than the [arc] that is in correct ratio to *KZ*, just as it is less than *ΘT*.

In these respects, then, this method is superior to the former. But it might be inferior to the other with respect to the ease of making the map, since [in the former method] it was possible to inscribe each locality by revolving and moving the ruler from side to side, with just one of the parallels drawn and divided [into degrees]; whereas here such [a ruler] is of no advantage because of the bending of the meridian lines from the central [meridian], so that all the circles have to be drawn on the map, and positions falling between the grid lines have to be guessed at by calculating on the basis of the recorded fractional parts with reference to the [four] whole sides [of the grid] that surround [the place in question]. Even so, I think that, here as on all occasions, the superior and more troublesome [method] is to be preferred to the inferior and easier one; but all the same, one should hold on to descriptions of both methods, for the sake of those who will be attracted to the handier one of them because it is easy.[82]

[82]In some manuscripts (including **U** but not **X**) the end of Book 1 is followed by a list of the ratios between the parallels through Meroē, Soēnē, Rhodes, and Thulē, and the equator. These are simply extracted from the foregoing chapter.

# Book 2

## 1. Preface to the detailed guide

Let this be the end of our outline of the general assumptions about world cartography and the revision of the map that would be in accordance with [1] up-to-date research on the known parts of the earth (that is, our *oikoumenē*), and [2] both the correct proportion of the places with respect to each other and the greatest possible similarity in shape [to the real *oikoumenē*], and [3] the variety of map.

Here we shall begin the detailed guide. But we first make the following observation: the numbers of degrees in longitude and latitude of well-trodden places are to be considered as quite close to the truth because more or less consistent accounts of them have been passed down without interruption; but [the coordinates] of the [places] that have not been so traveled, because of the sparseness and uncertainty of the research, have been estimated according to their proximity to the more trustworthily determined positions or relative configurations, so that none of the [places] that are to be included to make the *oikoumenē* complete will lack a defined position. We have therefore put the degrees corresponding to each [place] at the outer edge of the columns in the manner of a table, setting the [degrees] of longitude before those of latitude, so that if anyone should come across corrections from fuller research, it will be possible to put them alongside in the remaining spaces of the columns.

We have chosen an order [of presentation] with forethought to convenience in the drawing of the map in every respect, namely progressing toward the right, with the hand proceeding from the things that have already been inscribed to those that have not yet [been inscribed]: this would be achieved by having the more northern [places] drawn before the more southerly ones, and the more western before the more eastern, because our convention is that "up" with respect to the map-makers' or spectators' view means "north," and "right" means "east" in the *oikoumenē*, both on a globe and on a planar map. For this reason we have first set down the localities in Europe, which we, too,[1] have

[1]In defining the extent of the three continents the division between Africa and Europe is more or less clearcut, but that between Europe and Asia is to a certain extent arbitrary. Ptolemy indicates here that he is following some of his predecessors, perhaps, most immediately, Marinos.

divided from Libyē by the Straits of Hēraklēs, and from Asia, first by the seas between [Europe and Asia] and by Lake Maiōtis, and thereafter by the river Tanais and the meridian from this to the unknown land [to the north]. Next, [we have set down] the localities in Libyē, dividing this, too, from Asia, first by the seas that extend from the bay near Cape Prason in Aithiopia to the Bay of Arabia, and from that point on by the isthmus from the inlet at Hērōopolis to the Mediterranean,[2] which separates Egypt from Arabia and Judea; this is both in order not to split Egypt by dividing [the continents] by the Nile, and also because it is better, when possible, to divide the continents by seas instead of rivers.[3] Last, we shall write down the [localities] in Asia.

We will keep to the same principles also in each continent with respect to its parts as [we do] for the whole world and the entire *oikoumenē* with respect to [the continents], that is, we will again begin by recording the more northern and western countries and the adjacent seas and islands and the more noteworthy things of each kind. We will distinguish these parts [of the continents] by the boundaries of the satrapies or provinces, making the guide, as we originally promised, only as detailed as will be useful for recognizing and including places [on the map], while leaving out the great mass of reports about the characteristics of the peoples (unless perhaps some bit of current knowledge calls for a brief and worthwhile note).[4]

Moreover, this method of exposition will also make it possible, for anyone who wishes, to draw the parts of the *oikoumenē* on planar surfaces, individually or in groups of provinces or satrapies, in whatever way they might fit the proportions of the maps.[5] The [localities] contained by each chart will then be inscribed at the appropriate scale and relative placement. In this [kind of map] it will not much matter if we make the lines for the meridians parallel, and the lines for the parallels [of latitude] straight, so long as the degree intervals on the meridians have the same ratio to those on the parallels as a great circle has to the parallel that is to be in the middle of this map.

Now that these principles have been set out, let us begin the detailed guide here.

*[As a specimen of Ptolemy's geographical catalogue, we translate the chapters covering the four provinces of Roman Gaul, along with the caption for the corresponding regional map in 8.5. Map 8, constructed from the text, portrays the outlines of the provinces and all the cities and towns listed by Ptolemy; it was*

[2]Literally, "our sea."

[3]Some ancient geographers (e.g., Strabo) treated the part of Egypt east of the Nile as part of Asia. Ptolemy does not mention here that his world map will also connect Libyē to Asia by the unknown land to the south of Cape Prason.

[4]This happens seldom, notably in the sections on southeast Libyē and India Beyond the Ganges.

[5]Ptolemy here anticipates his division of the world map into regional maps in Book 8.

*not possible to indicate all the tabulated physical features in a map this small, and the districts of the native peoples (only roughly delineated by Ptolemy) have been omitted.*]

## 7. *The disposition of Celtogalatia [Gallia] Aquitanica*[6]

Celtogalatia is divided into four provinces: Aquitania, Lugdunensis, Beltica [Belgica],[7] and Narbonensis.

[*Description of the province's boundary*]
Aquitania is bounded on the west by the Aquitanian Ocean with the following outline of coast:

| After Oiasso Promontory in the Pyrenees:[8] | | |
|---|---|---|
| Mouth of R. Aturis [*Adour*] | 16¾ | 44¾ |
| Mouth of R. Sigmatis [*Leyre*] | 17 | 45⅓ |
| Cape Curianum | 16½ | 46 |
| Mouth of R. Garuna [Garumna, *Garonne*] | 17½ | 46½ |
| The middle of its extent | 18 | 45⅓ |
| The river's source | 19½ | 44¼ |
| Harbor of the Santones | 16½ | 46¾ |
| Cape of the Santones | 16½ | 47¼ |
| Mouth of R. Canentellus [Carantonus, *Charente*] | 17¼ | 47¾ |
| Cape Pictonium | 17 | 48 |
| Sicor harbor | 17½ | 48¼ |
| Mouths of R. Liger [*Loire*] | 17⅔ | 48½ |

| On the north [Aquitania is bounded] by part of the province Lugdunensis along the aforesaid R. Liger as far as its turning to the south, which has the position | 20 | 48½ |
|---|---|---|

[6]In 2.7–10 we have attempted to give the Latin forms that underlie Ptolemy's Greek place names, making allowances for the difference of language. (For example, we tacitly restore the Latin termination *-sis* where Ptolemy has substituted the Greek *-sia*.) We have added in brackets some significantly different attested Latin forms, and, in italics, modern counterparts where appropriate.

[7]The manuscript tradition of the *Geography*, as well as the *Handy Tables*, is unanimous in giving "Beltikē" rather than the expected "Belgikē."

[8]The coordinates for this feature (15, 45⅚) were provided in 2.6 as the last point along the northern coast of Hispania Tarraconensis. In some manuscripts the locations of such "connecting points" are frequently repeated to spare the cartographer the inconvenience of referring back to earlier chapters.

| | | |
|---|---|---|
| The eastern side borders part of Lugdunensis along the R. Liger to its source, which has the location | 20 | 44½ |
| and [by part] of Narbonensis to its limit at the Pyrenees, which has the location | 19 | 43⅙ |

The southern side borders part of the Pyrenees as far as Narbonensis.[9]

*[Interior features]*

| | | |
|---|---|---|
| The Pictones occupy the most northern parts of Aquitania near the sea. Their cities [include]: | | |
| Ratiatum [*Rézé*] | 17⅚ | 48⅓ |
| Limonum [*Poitiers*] | 18 | 47⅚ |
| Below[10] them are the Santones and their city, | | |
| •Mediolanium[11] [*Saintes*] | 17⅔ | 46¾ |
| Below these are the Bituriges Vibisci [Vivisci] and their cities: | | |
| Noviomagus | 17⅔ | 46¼ |
| •Burdigala [*Bordeaux*] | 18 | 45½ |
| Below these and as far as the Pyrenees Mountains are the Tarbelli and their city, | | |
| Aquae Augustae [Aquae Tarbellicae, *Dax*] | 17 | 44⅔ |
| Inland and below the Pictones are the Limovici [Lemovices] and the city, | | |
| Augustoritum [*Limoges*] | 17⅔ | 47¾ |
| [Below] these are the Cadurci and the city, | | |
| Dueona [Divona, *Cahors*] | 18 | 47¼ |
| Below these are the Petrocorii[12] and the city, | | |
| Vesuna [Vesunna, *Périgueux*] | 19⅚ | 46⅚ |
| Alongside all of these to the east and also occupying the [land] beyond the R. Liger are the Bituriges Cubi, and the city, | | |
| Avaricum [*Bourges*] | 20¼ | 46⅔ |
| Again below the Petrocorii live the Nitiobriges and the city, | | |
| Aginnum [*Agen*] | 19⅚ | 46⅓ |

[9]The outline of the Pyrenees was described in 2.6 as following a curved line from Cape Oiasso through a "midpoint" (17, 43) to the Temple of Venus (20⅓, 42⅓).

[10]"Below" means, approximately, to the south, but Ptolemy applies the term very loosely.

[11]Cities marked with a dot are the Noteworthy Cities. In the manuscripts these are indicated either by a symbol preceding the name or by a large initial letter jutting into the left margin.

[12]The Petrocorii (and Vesunna) were in fact north of the Cadurci (and Cahors).

| | | |
|---|---|---|
| Below these are the Vasarii [Vasates or Vocates] and the city, | | |
| Cossium [Cassio Vasates, *Bazas*] | 18½ | 46 |
| Below these are the Tabali [Gabali] and the city, | | |
| Anderidum [Anderitum, *Javoulx*] | 19¾ | 45½ |
| And below the Tabali are the Datii and the city, | | |
| Tasta | 19 | 44¾ |
| Below these are the Auscii and the city, | | |
| Augusta [*Auch*] | 18 | 45 |
| East of these are part of the Arverni, among whom is the city, | | |
| Augustonemetum [*Clermont*] | 20 | 45 |
| And below the Auscii are the Vellauni [Vellavi], and their city, | | |
| Ruessium [Revessio, *Saint-Paulien*] | 18 | 44½ |
| Below these are the Rutani [Ruteni] and the city, | | |
| Segodunum [*Rodez*] | 17¾ | 43½ |
| Adjacent to the Pyrenees are the Cotveni [Convenae] and their city, | | |
| Lugdunum colonia [Lugdunum Convenarum, *Saint-Bertrant de Cominges*] | 17 | 44 1/2 |

## *8. The disposition of [Gallia] Lugdunensis*

[*Description of the province's boundary*]

The sides of Lugdunensis that neighbor Aquitania have been described. Of the remainder, the [side] to the west and [along] the next part of the Ocean has the following outline:

| | | |
|---|---|---|
| After the mouths of the R. Liger: | | |
| Brivates [harbor] | 17⅔ | 48¾ |
| Mouths of R. Herius | 17 | 49¼ |
| Vidana (harbor) [Vindana, *Port Navalo*] | 16½ | 49⅔ |
| Cape Gabaeum [Gobaeum] | 15¼ | 49¾ |

The side to the north and along the Britannic Ocean is as follows:

| | | |
|---|---|---|
| After Gabaeum Promentory: | | |
| Salioncanus (harbor) [Staliocanus] | 16½ | 50 |
| Mouths of R. Titus [Tetus] | 17⅓ | 50⅓ |

| | | |
|---|---|---|
| Biducesii[13] | | |
| Aregenua Biducesiorum [*Vieux*] | 18 | 50½ |
| Venelli [Unelli]: | | |
| Cruciatonnum [Crociatonum, *Carentan*] | 18⅚ | 50⅓ |
| Mouths of R. Olina [*Orne*] | 18¾ | 51 |
| Lexubii [Lexovii]: | | |
| Noeomagus (city) [Noviomagus, *Lisieux*] | 19½ | 51⅙ |
| Baleti [Caletes?]: | | |
| Mouths of R. Sequana [*Seine*] | 20 | 51½ |
| The eastern side borders Beltica along the R. Sequana, the middle of which occupies[14] | 24 | 47⅓ |
| And further in a straight line with [the river] as far as a terminus, which has the location | 25 | 45½ |
| The southern [side] is bounded from that point by part of Narbonensis to the aforesaid terminus with Aquitania along the Cemmena [Cebenna, *Cévennes*] Range, the middle of which [has the location] | 20⅓ | 44½ |
| [*Interior features*] | | |
| The Caletes occupy the northern coast from the R. Sequana on. Their city, | | |
| Iuliobona [*Lillebonne*] | 20¼ | 51⅓ |
| After them, the Lexubii; then the Venelli; after them the Biducesii; and lastly, as far as Gabaeum Promentory, the Osismii, whose city, | | |
| Vorganium | 17⅔ | 50⅙ |
| The Veneti [occupy] the western coast, below the Osismii. Their city, | | |
| Darioritum [Dariorigum, *Vannes*] | 17⅓ | 49¼ |
| Below them, the Samnitae [Namnetes?] along the banks of the R. Liger. | | |

[13]As in the sections describing the inland features, Ptolemy here first lists the people inhabiting a particular district and then the names and coordinates of their cities.

[14]The manuscripts agree in calling this the midpoint of the river, but the point defined by these coordinates is in fact quite near the source, near Troyes (= Augustobona). Since the manuscripts give no source for the river (contrary to Ptolemy's normal practice when he gives a midpoint), it seems likely that the coordinates of the midpoint have dropped out of the text, and those given here pertain to the source.

| | | |
|---|---|---|
| Inland and east of the Veneti are the Aulircii Diablites [Aulerci Diablintes], and their city, | | |
| Noeodunum [Noviodunum, *Jublains*] | 18 | 50 |
| After them, the Arubii [Arvii] and their city, | | |
| Vagoritum | 18⅔ | 50 |
| After them and as far as the R. Sequana, are the Veliocasii [Velliocasses], and their city, | | |
| Ratomagus [*Rouen*] | 20⅙ | 50⅓ |
| And again, to the east of the Samnitae are the Andicavi [Andecavi or Andes] and their city, | | |
| Iuliomagus [*Angers*] | 18⅚ | 49 |
| Next after them are the Aulircii Cenomanni [Aulerci Cenomani], and their city, | | |
| Vindinum [Suindinum, *le Mans*] | 20¾ | 49⅓ |
| After them, the Namnetes and their city, | | |
| Condevincum [Condivicnum] | 21¼ | 50⅔ |
| Then as far as the R. Sequana are the Abrincatui and their city, | | |
| Ingena [*Avranches*] | 21¾ | 50¾ |
| The Aulircii [Aulerci] Eburovices follow below all the aforesaid [peoples] from the R. Liger to the R. Sequana. Their city, | | |
| Mediolanium [Mediolanum] | 20⅔ | 48 |
| Below them and along the R. Liger are the Riedones and their city, | | |
| Condate [*Rennes*] | 20⅔ | 47⅓ |
| And east of them are Senones, and their city, | | |
| Agedicum [Agedincum, *Sens*] | 21¼ | 47⅙ |
| Along the R. Sequana are Carnutes and their cities: | | |
| Autricum [*Chartres*] | 21⅔ | 48¼ |
| Cenabum [Genabum, *Orléans*] | 22 | 47⅚ |
| Below them are Parisii and their city, | | |
| Lucotecia [Lutetia, *Paris*] | 23½ | 48½ |
| Below them are Tricasii [Tricasses] and their city, | | |
| Augustobona [*Troyes*] | 23¼ | 47¾ |
| Again below the aforesaid peoples and along the R. Liger are to be found the Turupii [Turones] and their city, | | |
| Caesarodunum [*Tours*] | 20¾ | 46½ |
| Below them, and bordering the Arverni who live along the Cemmena Range, are Segusianti | | |

| | | |
|---|---|---|
| [Segusiavi] and their cities: | | |
| Rodumna [*Roanne*] | 21 | 45⅚ |
| Forum Segusiantorum [F. Segusiavorum, *Feurs*] | 21½ | 45½ |
| East of the aforesaid [peoples] are Meldae [Meldi] and their city, | | |
| Iatinum | 23 | 47½ |
| After them and toward Beltica are Vadicassii and their city, | | |
| Noeomagus | 24⅓ | 46⅙ |
| East of the Arverni and as far as the northward branch of the R. Rhodanus are the people of the Aedui and their cities: | | |
| •Augustodunum [*Autun*] | 23⅔ | 46½ |
| Cabullinum [Cabillonum, *Chalon-sur-Saône*] | 22⅚ | 45⅔ |
| •Lugdunum Metropolis [*Lyon*] | 23¼ | 45⅓ |

## *9. The disposition of [Gallia] Beltica*

[*Description of the province's boundary*]
The western side of [Gallia] Beltica along Lugdunensis has been described. The northern, along the Britannic Ocean, is as follows:

| | | |
|---|---|---|
| After the mouths of the R. Sequana: | | |
| Mouths of the R. Phrudis | 21¾ | 52⅓ |
| Cape Itium | 22¼ | 53½ |
| Morini: | | |
| •Gesoriacum (port) | 22½ | 53½ |
| Mouths of R. Tabula [Tabullas or Tabuda] | 23½ | 53½ |
| Mouths of R. Mosa [*Meuse*] | 24⅔ | 53½ |
| Batavi: | | |
| Lugodinum [Lugdunum, *Leyden*] | 26½ | 53⅓ |
| Western mouth of R. Rhenus [*Rhine*] | 26¾ | 53⅓ |
| Middle mouth of the river | 27 | 53⅙ |
| Eastern mouth of the river | 27⅓ | 54 |

| | | |
|---|---|---|
| The eastern side is bounded by the aforesaid river along Great Germania. The [river's] source is located at | 29⅓ | 46 |
| The westward branch at the R. Obrinca [Obringa] | 28 | 50 |

And [after the source of the Rhenus the eastern side

| | | |
|---|---|---|
| is bounded] by the mountains from the source to the Alps, called the Adulas [Adula] Range | 29½ | 45¼ |
| The Iurassus [*Jura*] range[15] | 26¼ | 46 |
| | | |
| The southern side neighbors the remaining part of Gallia Narbonensis, and extends from the aforesaid common terminus of Lugdunensis and Narbonensis to the common endpoint of the Alps and the Adulas Range, which occupies | 29½ | 45¼ |
| | | |
| [*Interior features*] | | |
| The Atribatii [Atrebati] occupy the coast, and also hold the inland along the R. Sequana. Their city, Metacum [Nemetacum, *Arras*] | 22 | 51 |
| After them and to the east are the Bellovaci, and their city, Caesaromagus [*Beauvais*] | 22½ | 51⅓ |
| After them likewise the Ambiani, and their city, Samarobriva [*Amiens*] | 22¼ | 52½ |
| After them, the Morini, and their inland city, Tarvanna [Tarvenna, *Thérouane*] | 23⅓ | 52⅚ |
| Then, after the R. Tabula, the Tungri and their city, Atuacutum [Atuatuca or Aduatuca, *Tongres*] | 24½ | 52⅚ |
| Then after the R. Mosa, the Menapii and their city, Castellum [*Cassel*] | 25 | 52¼ |
| Below these peoples live, furthest north, the Nerusii [Nervii] and their city, Bagacum [*Bavay*] | 25¼ | 51⅔ |
| Below them the Subanectes [Silvanectes], and their city east of the R. Sequana, Ratomagus [Rotomagus] | 22½ | 50 |
| Below them the Romandues [Veromandui], and their city Augusta Romanduorum [A. Veromanduorum] | 25½ | 50 |
| Below them the Vessones [Suessiones], and their city east of the R. Sequana, Augusta Vessonum [A. Suessionum, *Soissons*] | 23½ | 48⅚ |
| After them along the river, the Remi and their city, •Durocottorum [Durocortorum, *Reims*] | 23⅚ | 48½ |

[15]These mountains are not part of the boundary.

| | | |
|---|---|---|
| East of the Remi and to the north are the Triberi [Treveri], and their city, | | |
| Augusta Triberorum [A. Treverorum, *Trèves*] | 26 | 48⅙ |
| To the south are the Mediomatrici, and their city, | | |
| Divodurum [*Metz*] | 25½ | 47⅓ |
| Below these and the Remi are Leuci and their cities: | | |
| Tullium [Tullum, *Toul*] | 26⅙ | 47 |
| Nasium [*Naix*] | 27⅚ | 46⅔ |
| Of the land along the R. Rhenus, that from the sea southwards to the R. Obrinca is called Lower Germania, in which are the cities (starting from the west of the R. Rhenus): | | |
| Batavi: | | |
| Botavodurum [Batavodurum] (inland) | 27¼ | 52⅙ |
| Below this, Veterra [Vetera] in which is Legion 30 "Ulpia" | 27½ | 51⅚ |
| Then Agrippinensis [*Cologne*] | 27⅔ | 51½ |
| Then Bonna [*Bonn*] Legion 1 "Minervia" | 27⅔ | 50⅚ |
| Then "Traiana" Legion 22 | 27½ | 50½ |
| Then Mocontiacum [Moguntiacum, *Mayence*] | 27⅓ | 50¼ |
| The [land] south of the R. Obrinca is called Upper Germania, in which are the cities (beginning from the R. Obrinca): | | |
| Nemetes: | | |
| Noeomagus [Noviomagus, *Spire*] | 27⅔ | 49⅚ |
| Rufiniana | 27⅔ | 49½ |
| Vangiones: | | |
| Borbetomagus [*Worms*] | 27⅚ | 48⅚ |
| Argentoratum [*Strasbourg*] Legion 8 "Augusta" | 27⅚ | 48¾ |
| Triboci: | | |
| Breucomagus [*Brumat*] | 27⅚ | 48⅓ |
| Elcebus [*Ell*] | 28 | 48 |
| Raurici [Rauraci]: | | |
| Augusta Rauricorum [A. Rauracorum, *Augst*] | 28 | 47½ |
| Argentovaria [Argentaria] | 27⅚ | 47⅔ |
| Below these and the Leuci live the Longones [Lingones], and their city, | | |
| Andomatunnum [Andomatunum, *Langres*] | 26¼ | 46⅓ |
| And beyond the range below them, which is called | | |

| | | |
|---|---|---|
| Iurassus, the Helvetii [live] along the R. Rhenus. | | |
| Their cities: | | |
| Ganodurum [*Soleure*] | 28½ | 46½ |
| Forum Tiberii | 28 | 46 |
| Below these are the Sequani, and their cities: | | |
| Diatavium | 25⅙ | 45⅔ |
| Visontium [Vesontio, *Besançon*] | 26 | 46 |
| Equestris [*Nyon*] | 27 | 45⅔ |
| Avanticum [Aventicum, *Avenche*] | 28 | 45½ |

## *10. The disposition of [Gallia] Narbonensis*

[*Description of the province's boundary*]

| | | |
|---|---|---|
| The sides of [Gallia] Narbonensis along the three adjacent provinces have been described. Of the rest, the side to the east is bounded by the western Alps Mountains from the Adulas Mountains to the mouths of the R. Varus [Var], the position of which [occupies] | 27½ | 43 |

The southern side of Narbonensis is bounded by the rest of the Pyrenees from Aquitania to the ridge at the Mediterranean Sea, on which is the Temple of Venus, and thereafter by the Sea of Gallia to the mouths of the R. Varus. The coast has the following outline:

| | | |
|---|---|---|
| After the Temple of Venus | 20⅓ | 42⅓ |
| Mouths of the R. Illeris [Illiberis, *Tech*] | 21 | 42⅔ |
| Mouths of the R. Ruscio [Ruscino, *Tet*] | 21¼ | 42¾ |
| Mouths of the R. Atax [*Aude*] | 21½ | 42¾ |
| Mouths of the R. Orobis [*Orb*] | 21¾ | 42¾ |
| Mouths of the R. Arauris [*Hérault*] | 22 | 42⅚ |
| Agatha (city) [*Agde*] | 22¼ | 42⅚ |
| Mount Setius [*Cette*] | 22½ | 42½ |
| Fossae Marianae | 22¾ | 42⅔ |
| Western mouth of the R. Rhodanus [*Rhone*] | 22⅚ | 42⅔ |
| Eastern mouth of the R. Rhodanus | 23 | 42⅚ |

[*Digression: description of the Rhone and its tributaries*]

The bend in the river below Lugdunum in the

| | | |
|---|---|---|
| direction of the Alps | 23 | 45¼ |
| The part of it at Lake Leemena [Lemanus, *Léman*] | 27¼ | 45 |
| The source of the river | 28⅓ | 44⅓ |
| Of the rivers that meet it, the Arar [*Saône*] and the Dubis [*Doubs*] flow into the part north of the city Lugdunum, after they have first joined together. | | |
| The sources of the R. Arar, which flows from the Alps Mountains, are at | 28⅔ | 44⅔ |
| The sources of the R. Dubis, which flows below [the R. Arar], are at | 28½ | 44½ |
| These also flow north from the Alps and turn west, and their point of meeting is at | 25⅓ | 45½ |
| The meeting [of the combined rivers] with the R. Rhodanus | 24 | 45½ |
| Into the part [of the Rhodanus] south of the city Vienna likewise flow the R. Isar [Isara, *Isère*] and the R. Druentia [*Durance*] from the Alps Mountains | | |
| The sources of the R. Isar | 28 | 44 |
| The source of the R. Druentis | 28 | 43¾ |
| Again the meeting of the R. Isar with the Rhodanus is at | 22⅔ | 44½ |
| Likewise the meeting of the R. Druentis [with the Rhodanus] occupies | 22⅔ | 43⅚ |

[*Continuation of the description of the coast*]

After the R. Rhodanus on the sea, there is situated

| | | |
|---|---|---|
| Avatili [Anatili]: | | |
| Maritima colonia (city) | 23½ | 43 1/12 |
| Then the mouths of the R. Caenus | 23¾ | 43 |
| Comani [Commoni]: | | |
| •Massalia [Massilia, *Marseilles*] (Greek city) | 24½ | 43 1/12 |
| Tauroentium | 24⅔ | 42⅚ |
| And Cape Citharistes | 25 | 42⅔ |
| Olbia (city) | 25⅙ | 42¾ |
| And the mouths of the R. Argenteus [*Argens*] | 25⅔ | 42¾ |
| And Forum Iulium colonia [Iulii colonia, *Fréjus*] | 26½ | 42⅚ |
| Decatii: | | |
| Antipolis [*Antibes*] | 27 | 43 |
| And the mouths of the R. Varus | 27½ | 43 |

[*Interior features*]

| | | |
|---|---|---|
| The most western part of Narbonensis is occupied by the Volcae Tectosaces [Tectosages]. Their inland cities: | | |
| Illiberis [*Elne*] | 19¾ | 43¼ |
| Ruscino [*Castel-Roussillon*] | 20 | 43½ |
| Tolosa colonia [*Toulouse*] | 20⅙ | 43¼ |
| Cessero [*Saint-Thiberi*] | 21¼ | 44 |
| Carcaso [*Carcassonne*] | 21 | 43¾ |
| Baetirae [*Béziers*] | 21½ | 43¼ |
| •Narbo colonia [*Narbonne*] | 21 | 43¼ |
| After them and as far as the R. Rhodanus are the Volcae Arecomii [Arecomici]. Their inland cities: | | |
| Vindomagus [*le Vigan*] | 21½ | 44½ |
| •Nemausus colonia [*Nîmes*] | 22 | 44½ |
| Then east of the R. Rhodanus are, farthest to the north, the Allobriges [Allobroges] below the Medulli,[16] and their inland city, | | |
| •Vienna [*Vienne*] | 23 | 45 |
| Below them and farthest to the west are the Sengalauni [Segalauni] and their city, | | |
| Valentia colonia [*Valence*] | 23 | 44⅓ |
| Farther east are the Tricastini and their city, | | |
| Noeomagus | 26½ | 45 |
| Then below the Sengalauni are the Cavari and their inland cities: | | |
| Acusio colonia [Acusiorum colonia] | 23 | 44⅔ |
| Avennio colonia [Avenio, *Avignon*] | 23 | 44 |
| Arausio [*Orange*] | 24 | 44½ |
| Cabellio colonia [*Cavaillon*] | 24 | 44 |
| And below them the Salves, and likewise their cities: | | |
| Tarusco [*Tarascon*] | 23 | 43⅔ |
| Glanum [*Saint-Remi*] | 23½ | 43½ |
| Arelatum colonia [*Arles*] | 22¾ | 43⅓ |
| Aquae Sextiae colonia [*Aix*] | 24½ | 43⅔ |
| Ernaginum [*Saint-Gabriel*] | 24 | 43¾ |
| Below the Tricastini are the Voconti and their city, | | |
| Vasio | 26 | 44½ |
| Below them are the Memini and their city, | | |

[16]The Medulli are not mentioned elsewhere in the text as we now have it.

| | | |
|---|---|---|
| Forum Neronis [*Carpentras*] | 25⅔ | 44¾ |
| Below them are the Elycoci and their city, | | |
| Albaugusta [Alba Augusta] | 26 | 43⅓ |
| East of the Voconti and the Memini are the Sentii and their city, | | |
| Dinia [*Digne*] | 27¾ | 44⅓ |

[*Islands in the Mediterranean*]

| | | |
|---|---|---|
| There are [the following] islands situated below Narbonensis: | | |
| Agatha [*Agde*] at the city of the same name; the island's position is | 22⅙ | 42⅙ |
| After it, Blasco | 22½ | 42⅙ |
| The five Stoechades below Citharistes; the position of their center is | 25 | 42¼ |
| Below the R. Varus is the island Lerona [*île Sainte-Marguerite*], whose position | 27¼ | 42¼ |

# *Book 7*

## [*Concluding paragraph of 7.4*]

Let [the foregoing] be the way that the guide of the individual provinces and satrapies of the *oikoumenē* is composed. But since we demonstrated at the beginning of the compilation how the known part of the earth could be mapped on a globe, and also on a plane surface, in a way that is, as far as possible, both similar [in appearance] and proportionate to the things that are comprehended on the solid globe, it is appropriate to add to these portrayals of the whole *oikoumenē* a summary caption that will indicate the things that are generally seen [in the map]. This, too, would be of appropriate dimensions if it were as follows:

## *5. Summary caption of the map of the* oikoumenē

**Our *oikoumenē* having been divided into three continents by the earlier [writers], who have made detailed and accurate researches and left records to the best of their ability of what each had at his disposal, we, too, having seen some things ourselves, and also having taken over other things accurately from [the earlier writers], have taken forethought to sketch, as it were, a sort of map of the entire *oikoumenē*, so that none of those things that are specifically useful and that together with written accounts can adorn the soul and incite it to perspicacity concerning nature should be unknown to lovers of knowledge.**[1]

**The part of the world [contained] in our *oikoumenē* is bounded to the east by the unknown land that is situated next to the eastern peoples of Great Asia, [namely] the Sinai and the people in Sērikē; to the south likewise by the unknown land that encloses the Sea of India and surrounds Aithiopia south of Libyē (this [part of Aithiopia] is called the country of Agisymba); to the west by both the unknown land surrounding the Aithiopian Bay of Libyē and the adjacent Western Ocean,**

[1]The foregoing paragraph appears elsewhere in **X**, and may be spurious. See Textual Notes (Appendix G). With or without the preface, Ptolemy's caption would be of "appropriate dimensions" for a map of monumental size.

**which lies next to the most western parts of Libyē and Europe; and to the north by the continuation of the Ocean that contains the islands of Britain and the most northern parts of Europe (this [sea] is called the Duēcalidonian and Sarmatian [Ocean]), and by the unknown land that is situated next to the most northern countries of Great Asia, [namely] Sarmatia, Skythia, and Sērikē.**

**These seas are contained by the *oikoumenē*:**

- **The Mediterranean, together with the bays that are connected to it, besides the Adriatic Bay and the Aegean Sea, the Propontis and Pontos, and Lake Maiōtis. It has an opening to the Ocean only[2] through the Straits of Hēraklēs, in the manner of a peninsula, making the strait, as it were, an isthmus of the sea.**
- **The Hyrkanian or Caspian Sea is enclosed on all sides by the land, like an island in reverse.**
- **Likewise the whole sea around the Sea of India, together with the bays connected to it, besides the Arabian and Persian Bays, the [Bay] of Ganges, and the Bay that is specifically called Great. This [sea], too, is contained by land on all sides.[3]**

**Hence of the three continents, Asia is connected to Libyē both by the land-strait at Arabia, which also separates the Mediterranean Sea from the Arabian Bay, and by the unknown land that surrounds the Sea of India. And [Asia] is connected to Europe by the land-strait between Lake Maiōtis and the Sarmatian Ocean where the river Tanais crosses through. Libyē is divided from Europe by just the Straits [of Hēraklēs], and is not in itself attached [to Europe] at all, but only by way of Asia, since [Asia] is attached to both of them, flanking them both on the east.**

**Moreover, Asia is the first of the continents in respect to size, Libyē second, Europe third. Likewise, of the seas that have been said to be contained by land, the Sea by India is again the first in size, the Mediterranean Sea second, and the Hyrkanian or Caspian third.**

**And among the more noteworthy bays:**

- **the first and greatest is the Bay of Ganges,**
- **the second the Persian Bay,**
- **the third the Great Bay,**

[2]This "only" is presumably intended to exclude the possibility (apparently advocated only in *Periplus* 64 (Casson 1989, 91) that Lake Maiōtis opens directly into the Ocean to the north. In the following sentence Ptolemy rejects the idea that the Caspian was open to the Ocean on the north, which had been common among earlier geographers (including the *Periplus*). See Casson 1989, 239–241.

[3]By enclosing the Indian Ocean with land on all sides, Ptolemy leaves the mainstream of Greek geographical tradition, which made it part of the all-encircling Ocean. See Introduction, p. 22.

— the fourth the Arabian Bay,
— the fifth the Aithiopian Bay,
— the sixth the Bay that is the Pontos,
— the seventh the bay that is the Aegean Sea,
— the eighth the bay that is Lake Maiōtis,
— the ninth the Adriatic,
— the tenth the bay that is the Propontis.

Among the more noteworthy islands or peninsulas:
— the first is Taprobanē,
— second Albiōn, one of the islands of Britain,
— third the Golden Peninsula,
— fourth Hibernia, one of the islands of Britain,
— fifth the Peloponnese,
— sixth Sicily,
— seventh Sardinia,
— eighth Corsica,
— ninth Crete,
— and tenth Cyprus.

The southern limit of the known world is bounded by the parallel 16⁵⁄₁₂° south of the equator, such that the great circle is 360°. This is the same number of degrees by which the parallel through Meroē is north of the equator. The northern limit is bounded by the parallel 63° north of the equator, and is drawn through the island of Thulē. Hence the known latitudinal dimension of [the *oikoumenē*] is 79⁵⁄₁₂°, or in round numbers, 80°; which is approximately 40,000 stades on the assumption that one degree contains 500 stades (as has been determined by the more accurate distance measurements), and that [therefore] the whole earth [contains] a perimeter of 180,000 stades.

Again, the eastern limit of the known world is bounded by the meridian drawn through the metropolis of the Sinai, which is 119½° on the equator, or approximately 8 equinoctial hours, east of the meridian drawn through Alexandria. The western limit [is bounded by] the meridian drawn through the Islands of the Blest, which is 60½°, or 4 equinoctial hours, from the meridian through Alexandria. [It is thus] 180° (i.e., a semicircle), or 12 equinoctial hours, from the most eastern meridian.

Hence the known longitudinal dimension of [the *oikoumenē*] amounts to:
— 90,000 stades on the segment along the equator, or,
— approximately 86,330⅓ stades on the segment along the most southern parallel of [the *oikoumenē*];

— **40,000 stades on the segment along the northernmost parallel; and again,**
— **approximately 72,000 stades on the parallel through Rhodes, along which the distance measurements have most often been made, and which is 36° from the equator; and**
— **82,336 stades on the parallel through Soēnē, which is 23⅚° from the equator, and situated about in the middle of the whole latitudinal dimension.**[4]

**[These figures are] according to the proportionality between the named parallels and the equator. Thus the longitudinal dimension of the *oikoumenē* is greater than the latitudinal dimension, in the most northern *klimata*[5] by approximately one-fiftieth of the latitudinal dimension;[6] in the latitude through Rhodes by about five-sixths of the latitudinal dimension; in the *klimata* through Soēnē by an equal amount to the latitudinal dimension plus approximately one-eighteenth; in the most southern parts by an equal amount plus approximately one-sixth, and in the parts under the equator by the same amount plus one-quarter.**

**The length of the longest day or night is, in the most southern of the named parallels, 13 equinoctial hours; on the parallel through Soēnē, 13½ hours; on the parallel through Rhodes, 14½ hours; on the most northern parallel through Thulē, 20 hours. Thus the difference over the whole latitudinal breadth makes 9 equinoctial hours.**[7]

[4]These stade figures are a curious mix of accurate and rough calculation. 90,000 stades along the equator follows directly from the assumption that one equatorial degree equals 500 stades. The number of stades along another parallel should equal the number of stades along the equator multiplied by the cosine of the latitude, a quantity that Ptolemy can compute by means of the table of chords in *Almagest* 1.11. The figures for the parallels through Rhodes and Thulē (the northernmost parallel) actually were obtained by applying simple ratios of 4:5 and 4:9 respectively to the figure for the equator. Accurate calculation would have led in these cases to 40,859 and 72,812 stades. For the parallels through Soēnē and at the southern limit of the *oikoumenē*, the chord table yields approximately 82,325 stades and 86,330⅔ stades, respectively.

[5]The term is used here in a broad sense of "latitudinal belts," not specifically the scheme of the seven *klimata*.

[6]In the foregoing section Ptolemy has rounded both the latitudinal dimension and the northernmost longitudinal dimension to the same figure, 40,000 stades. To get the excess of one-fiftieth stated here, Ptolemy must have compared a more accurately calculated figure for the longitudinal dimension with the rounded latitudinal dimension. The excess of fiv- sixths for the latitude of Rhodes again seems to have been derived from a longitudinal dimension more accurate than the round figure of 72,000 stades.

[7]The one-hour difference between the southern limit and the equator is, oddly, *added* to the eight-hour difference between the equator and Thulē. The actual difference in duration of longest day between Ptolemy's northern and southern limits is only seven hours, although there would be a nine-hour difference in the length of day on the northern hemisphere's summer solstice.

## 6. *The mapping of a ringed globe with the* oikoumenē

The caption for the general exposition [of the world map] would be of suitable length if carried out to this extent. It would not be out of place, however, to add a method of drawing the hemisphere containing the *oikoumenē* on a plane as it is seen [on a globe], and surrounded by a ringed globe; for many people have attempted such a demonstration, but it is manifest that they have attempted it in a most illogical fashion.

Let it be required to map on a plane a ringed globe[8] containing a part of the earth within itself, on the hypothesis that the eye occupies such a position that it will be collinear with the points where the meridian through the tropic points [of the ecliptic ring] intersects the parallel drawn through Soēnē on [the surface of] the earth. The [meridian] that bisects the longitudinal dimension of our *oikoumenē* will also be assumed to be lying under this [meridian through the tropic points]; and [the parallel through Soēnē] also approximately bisects the latitudinal dimension of the *oikoumenē*.[9]

Let the ratios of the sizes of the ringed globe and the earth and the distance of the eye be such that the whole known part of the earth can be seen in the space between the ring on the equator and [the ring] on the Summer Tropic, and let the southern semicircle of the ecliptic circle stand over the earth so that, when the *oikoumenē* has been placed in the northern hemisphere, it will not be hidden by [the ecliptic ring].

It is obvious at once on these assumptions [1] that the foregoing meridians [i.e., the meridian through the tropic points and the meridian bisecting the *oikoumenē*] will produce an illusion of a single straight line coinciding with the [earth's] axis, since the eye is situated in the plane through them, and also [2] that the parallel through Soēnē [will appear as a straight line] at right angles to [the central meridian] for the same reason. The other circles that are included [in the drawing] will appear curved with their concavities toward the straight lines, the meridians [concave] toward the [axis] through the poles, and the parallels [concave] toward the [parallel] through Soēnē; and the further away [the circles] are on either side, the more [curved] they will appear.

[8]The seven rings of the armillary sphere surrounding the globe (as described below) represent the equator, the ecliptic, the meridian through the equinoctial points, the two tropic circles, and the arctic and antarctic circles. Only the ecliptic and meridian would be of much use as movable rings surrounding a globe, since the other circles always stand over the same terrestrial locations.

[9]The hypothetical eyepoint is not the same as for the second map projection in 1.24. There, the viewer was assumed to be directly above the midpoint of the portion of the parallel through Soēnē contained within the *oikoumenē*, and far enough away so that the visible part of the globe is effectively a hemisphere. Here, the viewer is *in the plane* of the parallel through Soēnē, and near enough so that the entire *oikoumenē* can be seen inside the gap between the equatorial and Summer Tropic rings.

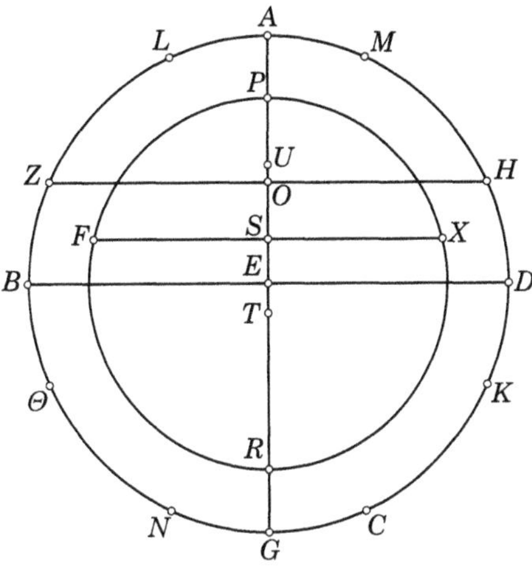


Fig. 18.

The procedure of carrying out the mapping in as close agreement as possible with the rules of optics will be easy for us [if carried out] as follows.[10]

[*Stage 1: Determination of a suitable size for the terrestrial globe relative to the rings.*] Let the meridian through the equinoctial points on the ringed globe be *ABGD* about center *E* and diameter *AEG*, with *A* imagined to be at the north pole, and *G* at the south pole [Fig. 18]. And let arcs *BZ*, *DH*, *BΘ*, and *DK* be taken at the intervals of the tropics from the equator, and *AL*, *AM*, *GN*, and *GC* at the [intervals] of the arctic and antarctic [circles] from the poles; and let the diameter of the summer [tropic] cut *AE* at *O*.

Now the parallel through Soēnē must be placed between *E* and *O*. [We know that] the ratio to a quadrant of the arc between the parallel through Soēnē and the equator is approximately 4 to 15, while the [ratio] of half *EO* to *EA* is approximately [that of] the same [number], 4, to 20.[11] We will therefore let *EA* be

[10]Ptolemy presents the whole of the following complex geometrical construction with reference to the single diagram Figure 20. For clarity we have added Figures 18–19 to show the intermediate stages of the construction.

[11]Ptolemy does not adequately explain the reasons for the following operations. He is going to draw a circle to represent the terrestrial globe inside the ring *ABGD*, and a map of the *oikoumenē* will be drawn inside this circle. The part of line *AG* that is inside the terrestrial circle will represent the meridian that bisects the *oikoumenē*. Distances along this line are intended to be proportional to arcs along the actual meridian, so that the center of the circle, *E*, will represent a point on the terrestrial equator, and a point (to be labeled *S*) that is about 4/15 of the circle's radius above the center will represent a point on the parallel through Soēnē. The hypothesis about how the eye is imagined to be viewing the ringed globe implies that *S* will be seen somewhere between *E* and *O*, and Ptolemy provisionally puts *S* approximately halfway between them. He knows that *EO*/*EA* (i.e., sin 23⅚°) is about 4/10, hence *ES*/*EA* is about 4/20, and the radius of the terrestrial circle will therefore be about ¾ *EA*. He now makes this radius (*EP*) *exactly* ¾ *EA*, and a bit later determines the definitive position of *S* using the more precise ratio *ES*:*EP* = 23⅚:90. All this seems needlessly roundabout.

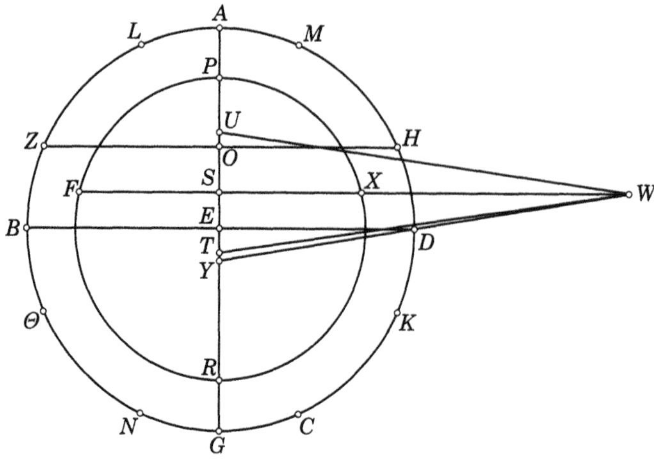


FIG. 19.

$^4/_3$ of the earth's radius. Hence let *EP* be taken as three of such [units] as *EA* is four. And with center *E* and radius *EP*, let circle *PR* be drawn [to represent] the circle circumscribing the earth in the same plane.

[*Stage 2: Construction of the midpoints of the principal parallels on the terrestrial globe.*] Let some straight line equal to *EP* be divided in the 90 equal divisions of one quadrant, and let *ES* be taken as 23⅚ divisions, *ET* as 16$^5/_{12}$ divisions, and *EU* as 63 of the same [divisions]. And let *FSX* be drawn across *EP* at right angles; it will fall of course along the parallel through Soēnē. Thus *T* will be [the point] through which is drawn the parallel bounding the southern limit of the *oikoumenē* and oppositely situated to the [parallel] through Meroē; and *U* will be the point through which is drawn the parallel bounding the northern limit and drawn through Thulē.[12]

[*Stage 3: Determination of a suitable eyepoint, and projection of the midpoints of the rings.*] Let some point *Y* be taken slightly to the south of *T*, and let *YD* be joined, and let *SX* and *YD* produced meet in *W* [Fig. 19].[13] Now if we imagine the circles that have been drawn as being in the plane through the tropic points and the poles [of the equator], and the eye as being at *W*, then by

[12]As remarked in the preceding note, the midpoints of the parallels are determined not by linear perspective, but according to a similar hypothesis to the map projections in 1.24, that latitudinal intervals along the central meridian should be represented by proportional linear distances on the map. At this point we can draw only one of the parallels, that through Soēnē, because we know that it must appear as a straight line to a viewer in its plane.

[13]The eye must occupy a suitable position on the extension of the plane of *FSX* so that everything on the globe between *U* and *T* will be visible between the rings, i.e., between *H* and *D*. Ptolemy knows, but does not point out, that if the eye is near enough to see *T* above *D*, it will also see *U* below *H*, but not necessarily vice versa. For this reason he finds an eyepoint *W* such that *D* is in line with a point just below *T*.

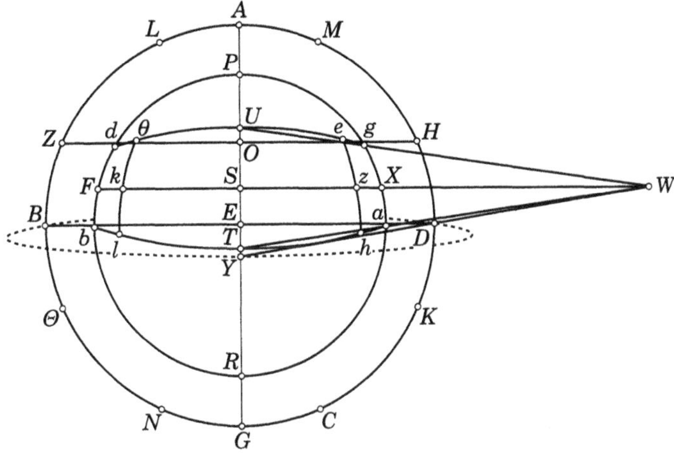


FIG. 20.

our assumptions, straight lines drawn from $W$ through $M, H, D, K$, and $C$ to $AG$ will make the points of intersection on [$AG$] through which we will draw the segments of the five parallels [of the ringed globe] that are nearer the eye: for example, $Y$ [is the point] through which we will draw the [part] of the [celestial] equator near $D$.[14] And the [lines] drawn from $W$ to $L$, $Z$, $B$, $\Theta$, and $N$ will intersect $AG$ in the points through which we will draw the segments of the same parallels on the other side of the earth.[15]

[*Stage 4: Construction of the arcs representing the parallels of latitude.*] Similarly [Fig. 20], if we take on $PR$ the respective distances from the equator of the parallels to be drawn on the earth, such as $U$ and $T$, and the points of intersection that result from the straight lines joined to them from $W$ on the semicircle $PXR$, and the [points] opposite to these along parallel directions, we will get [the points] through which the segments of the parallels mentioned above will be drawn, for example $aTb$ and $gUd$.[16]

[14]This is a correct determination by linear perspective of the projections of the midpoints of the rings upon the axis $AG$ of the globe. While we are constructing these points, the plane of the diagram is assumed to be the plane through this axis and the eyepoint. When the picture itself is executed, we will be imagining the plane of the diagram as rotated around the axis 90°, as if our eye occupies point $W$. In order not to clutter the figure, Ptolemy illustrates the construction of the midpoints with only one specific instance.

[15]Most of these points will be concealed in the final picture, since they are behind the terrestrial globe. They are needed, however, to obtain the appropriate oval to represent the rings.

[16]These constructions are *not* true linear perspective, as the use of the eyepoint as a center of projection might suggest. Ptolemy's method does assure that the circular arcs representing the terrestrial parallels will have increasing curvature as they get further from the parallel through Soēnē, which is a qualitatively correct property of the appearance of the parallels on the globe.

[*Stage 5: Construction of the arcs representing the meridians.*] We then take on these the distances of the meridians that are going to be inserted on either side of *TU*, and also [cut them off] on straight line *FX*, in the ratios that pertain to the three parallels; and through [each] set of three points corresponding to a single ratio we draw [circular] segments for the meridians in question, e.g., *ezh* and *θkl* for the meridians that bound the longitudinal dimension. We will draw as many [circles] on the earth as is appropriate to the size of the map.

When we add the rings [to the picture], we must take care that each passes through the four determined points, in an oval shape and not one that terminates in a cusp at the points of intersection on the outermost circle, so that it does not give the illusion of a fracture, but rather it should take on a curvature even at this point that is comparable to the adjacent [curvature], even if the bends that form the ends of the ellipse fall outside the circle that surrounds the figure; for this is seen to occur in real [rings], too.[17]

Care must also be taken [1] that the circles are not mere lines, but have some suitable breadth and distinguishing color, and moreover [2] that the segments on the other side of the earth present more shaded colors than those that are toward the eye, and [3] that when the segments intersect, those farther from the eye are cut off by the nearer ones in agreement with the obstructions [that occur] in real [ringed globes]—this applying to both the rings and the earth—and [4] that the southern semicircle of the ecliptic [ring], the one that goes through the Winter Tropic [point], passes in front of the earth and that its northern semicircle, the one that [passes] through the Summer Tropic [point], is cut off by [the earth].

We will inscribe the appropriate names around these [rings] in the proper places; and also for the circles on the earth, the numbers for the distances and hours that were demonstrated [in 1.23] in the mapping of the *oikoumenē*; and around the outer circle, the names of the winds on the five established parallels and the poles, in accordance with the markings on the ringed globe.

[17]What Ptolemy means is illustrated by the equatorial ring which we have added to Figure 20 as a broken line (it is not drawn in the manuscripts). The ellipse that passes through *D*, *Y*, *B*, and the projection of *B* on *AG* has its major axis below and longer than *BD*. Ptolemy knows that it would be optically incorrect to portray the ring as a lozenge with sharp points at *B* and *D*, as a naive artist might do. The other rings that would be drawn in the complete picture include the arctic and antarctic circles, the two tropics, and the ecliptic (a meridian ring is already represented by the circle *ABGD*).

## *7. Caption for the flattening [of the* oikoumenē*]*

There will be [the following] caption for this flattening[18] [of the *oikoumenē*] too, which is both fitting and concise:

**This mapping of the ringed globe in a plane, with the earth enclosed, is assumed to occupy the position at which the eye is in a straight line with the intersections of the meridian through the tropic points (under which also lies the [meridian] bisecting the longitudinal dimension of our *oikoumenē*) and of the parallel drawn through Soēnē on the [surface of the] earth (which also approximately bisects the latitudinal dimension of the *oikoumenē*).**

**The ratios of the sizes of the globe and the earth and the distance to the eye are such that the whole known part of the earth can be seen in the gap between the [ring] along the equator and the [ring] along the Summer Tropic; and the southern semicircle of the ring through the middle of the zodiacal signs stands over the earth, so that it, too, does not obstruct the *oikoumenē* (which is placed in the northern hemisphere).[19] Thus the meridians mentioned above produce an illusion of a single straight line along the axis itself, since the eye is situated in the plane through them, and for the same reason the parallel through Soēnē appears [as a straight line] at right angles to the other line. But the rest of the inscribed circles appear curved, with their concavities toward the straight lines (that is, the meridians are [concave] toward the [meridian] through the poles, and the parallels toward the [parallel] through Soēnē). And the [circles] that are farther away on either side [appear] the more [curved]; for example, the arctic [circle] bulges more to the north than the Summer Tropic, and the Winter Tropic more to the south than the equator, and the antarctic [circle] still more than the Winter Tropic.**

**The known part of the earth has been laid out as having the Ocean in no wise flowing around it, but rather [the Ocean] borders only the boundaries of Libyē and Europe that are drawn in the directions of [the winds] *Iapyx* and *Thraskias*, in agreement with the researches of the more ancient [writers].**

[18]The Greek words used here and in the heading are more normally applied to the spreading out or flattening of a cloth or net. Here it appears to describe a perspective picture.

[19]This is not strictly true, since part of the *oikoumenē* is south of the equator.

# Book 8

## 1. *On the basis for dividing the* oikoumenē *into the [regional] maps*

[The foregoing] suffices, in my opinion, for all the things that ought to be incorporated in the guide to world cartography, both in the way of verifying more thoroughly [the reports of] those who have explored the countries that are distant from us, and in the way of providing a more convenient and at the same time more appropriate approach to making the maps. I imagine, after all, that it would be ridiculous to go on chapter after chapter, as our predecessors have done, saying through which places each of the parallels and meridians included in the map is to be drawn, since absolutely all the places, even those that are not situated on the circles that are displayed [in the map], have beside them the positions of the parallels and meridians that are drawn through them.

Since we have shown the caption for the *oikoumenē* that would be appropriate when the whole [of the *oikoumenē*] is contained in a single map, the next thing is to set out what the concise summaries will be if we divide it into many maps, so that all the catalogued [localities] can be inscribed while still being at an appropriate scale for clarity. For in the case of an undivided map, because of the need to preserve the ratios of the parts of the *oikoumenē* to each other, some parts inevitably become crowded together because the things to be included are near each other, and others go to waste because of a lack of things to be inscribed. In trying to avoid this, most [map-makers] have frequently been constrained by [the shapes and sizes of] the [planar] surfaces themselves to distort both the measures and the shapes of countries, as if they were not guided by their research. This is the case, for example, with all those who have given the greatest part of the map in the longitudinal and latitudinal dimensions to Europe (because the things inscribed [there] are so numerous and close together) and [who have given] the least [part] in longitudinal dimension to Asia and in latitudinal dimension to Libyē for the converse reason. This is [also] why they make the Sea of India turn northward after Taprobanē, because [the edge of] the [planar] surface blocked their continuing eastward, whereas there was nothing [else] to inscribe in the [part of] Skythia above.[1] Again, they made the Western Ocean turn away to the east [at its southern end] because [the edge of] the [planar] surface blocked them in the southern direction, and there, too, neither the bottom of Inner Libyē nor [that] of India had anything as they continued [southward beyond the known parts] that could be inscribed on the western coast [of Libyē and India].[2] And it is for such reasons as these that the doctrine that the Ocean flows around the whole world has arisen out of errors of drawing, to be turned [subsequently] into a confused narrative.

By such a division into [regional] maps we can escape the described effect, if we make the division in such a way that the countries that have more things in them take up a map either singly or in small groups and with greater spaces between the [parallel and meridian] circles, while those [countries] that have less in them and that are not entirely covered [by the guide] are contained in one map together with many similar [countries] and with smaller spaces between the circles. For it is no longer necessary to have all the maps in proper proportion to each other, but only to have the things in each [map] preserve their ratios to each other; just as, when we portray just a head, [it is necessary to preserve the ratios of] the parts of the head only, or, if it be just a hand, the parts of the hand only, but not the parts of the head with respect to the parts of the hand unless we [portray] the whole person in one figure. But just as there is no reason not to make the whole [portrait] in one case bigger, in another case smaller, so [there is no reason not to] make some of the parts bigger, and some of them smaller, when they are by themselves, in accordance with the room available on the surfaces provided.

It will not be very inaccurate, as we said at the beginning of the compilation, if we inscribe straight lines in place of the [meridian and parallel] circles for the regional maps at least, and if moreover the meridians are [drawn as] not converging, but also parallel to one another. For in the case of the whole *oikoumenē*, the limits of the latitudinal and longitudinal dimensions, because they were taken at great intervals, make the distortions in the extreme circles

[1]That is, Ptolemy asserts, the early map-makers had no information about places in northeast Asia, so in order not to draw the coastline beyond the edge of their maps, they drew the Indian Ocean as turning north immediately after the known part of the coast.

[2]Ptolemy apparently thinks that earlier cartographers shared his belief that the outline of the *oikoumenē* was approximately as he portrayed it, with the southern coastline of Asia (along the "Sea of India") eventually turning south to join the east coast of Libyē, enclosing what we know as the Indian Ocean. However, according to Ptolemy, the cartographers set the physical limits of the map, not where they believed the land mass to end, but where the place names ran out: in the case of Asia, just beyond Taprobanē, and in the case of Libyē, perhaps about the equator. Hence, in order to show somehow that the west coast of Libyē and the south coast of Asia extended further, they were forced to draw them as following the edges of the map, contrary to their own beliefs. (A specimen of a map drawn on such principles may be found on the frontispiece of volume 1 of the Loeb edition of Strabo.)

significant [if they are drawn as parallel straight lines], but in the case of each of the [regional] maps this is no longer so. Hence we said that the division into degrees should be made according to the ratio of the parallel that bisects [the latitudinal dimension of] the map to the great circle, so that we will fail to take account, not of the defect [accrued] over the entire dimension of the map, but only that over the [interval] from the middle to either boundary of the maps.

## *2. Which things are appropriate to include in the caption for each map*

Starting from such a basis for the division, we have made ten maps of Europe, four maps of Libyē, and twelve maps of the whole of Asia. We have set out the captions for each, putting first the continent to which the map belongs, its ordinal number, what countries it contains, approximately what ratio the parallel through its middle has to the meridian, and what the boundary of the whole map is. We have put below [this information] the elevations [of the pole] for the principal cities in each country, converted into the lengths of the longest days [that occur] there; and their longitudinal positions [converted] approximately into intervals from the meridian through Alexandria, whether to the east or to the west, in units of equinoctial hours; and for those that the ecliptic stands over, [we have recorded] whether the sun passes through the zenith once or twice [in a year], and how [the sun] is situated [on the ecliptic] with respect to the tropic points [when this happens].

We would also have added which fixed stars pass through the zenith [at each locality] if these actually kept constant latitudes with respect to the equator, that is, if they traveled always on the same [celestial] parallels. But we have shown in the *Mathematical Compilation* [i.e., the *Almagest*] that the sphere of the fixed stars shifts in the direction of the trailing [parts] of the heavens with respect to the tropic and equinoctial points, and [it does this] not about the poles of the equator, but about those of the ecliptic, just like the [spheres] of the planets.[3] For this reason, the same stars cannot culminate always for the same localities, but must shift, some of them to more northerly positions than before, others to more southerly ones. Therefore it seemed to us superfluous to make such a supplement to the caption, since using the star globe that we made in accordance with this theory [of precession] we can establish the position [of the sphere of the fixed stars] at the times in question with respect to the circle

[3]Ptolemy demonstrates this in *Almagest* 7.3. According to Ptolemy, Hipparchus, the discoverer of precession, already suspected that the precessional motion took place about the poles of the ecliptic, and not the equatorial poles; but the consequence of this, that the declinations of the fixed stars change gradually over time, had apparently not been noticed by geographical writers before Ptolemy. For Ptolemy's neglect of this effect in his criticism of Marinos, see p. 65 n. 23.

through the poles of both [the equator and ecliptic], and, revolving the whole [globe] along the graduated edge of the fixed meridian [ring], [we can] determine what point on [the ring] is as many degrees from the equator as the parallel through the place in question is in the same direction, and [so we can] conveniently find out whether no star at all passes through that point, or if there are many, and, [if so,] which one or ones.[4]

After these preliminary explanations, let us here begin the rest of what we have proposed.

*[We provide as a specimen of Ptolemy's regional map captions the chapter corresponding to the description of Gaul in 2.7–10.]*

## *5. Map 3 of Europe*

**The third map of Europe contains Gallia comprising four provinces, with the adjacent islands. The parallel through its middle has the ratio of 2:3 to the meridian. The map is bounded on the east by Italia, Rhaetia, and Germania; on the south, by the Sea of Gallia; on the west, by the Pyrenees Range and the Bay of Aquitania; and on the north, by the Britannic Ocean.**

**In Aquitania:**

**— Mediolanium has a longest day of 15¾ equinoctial hours, and is 2⅚ equinoctial hours west of Alexandria.**[5]

**— Burdigala has a longest day of 15½ equinoctial hours, and is 2⅚ equinoctial hours west of Alexandria.**

**In Lugdunensis:**

**— Augustodunum has a longest day of 15⅔ equinoctial hours, and is 2½ equinoctial hours west of Alexandria.**

**— Lugdunum has a longest day of 15½ equinoctial hours, and is 2½ equinoctial hours west of Alexandria.**

**In Beltica:**

**— Gesoriacum has a longest day of 16⅚ equinoctial hours, and is 2½ equinoctial hours west of Alexandria.**

**— Durocottorum has a longest day of 16 equinoctial hours, and is 2½ equinoctial hours west of Alexandria.**

[4]The construction of the star globe is described in *Almagest* 8.3.

[5]Like the other infomation in this chapter, these coordinates in hours are, in principle, derived from the information in degrees in 2.7–10.

**In Narbonensis:**

- **Massalia has a longest day of 15¼ equinoctial hours, and is 2⅖ equinoctial hours west of Alexandria.**
- **Narbo has a longest day of 15¼ equinoctial hours, and is 2 7/12 equinoctial hours west of Alexandria.**
- **Vienna has a longest day of 15½ equinoctial hours, and is 2½ equinoctial hours west of Alexandria.**
- **Nemausus has a longest day of 15 5/12 equinoctial hours, and is 2½ equinoctial hours west of Alexandria.**

## PLATES AND MAPS

# *Notes on the Plates and Maps*

*Plate 1.* Map of the world in Ptolemy's first projection, from Urb. gr. 82 (**U**), ff. 60v (59v)–61r (60r). This manuscript, which was made about A.D. 1300, is among the oldest Greek manuscripts of the *Geography* containing maps. See also pp. 43–44. © Biblioteca Apostolica Vaticana.

*Plate 2.* Map of the world in Ptolemy's second projection, from Urb. lat. 274, ff. 74v–75r. This manuscript is one of the numerous copies of the *Geography* produced by Nicolaus Germanus, and was dedicated by him to Pope Paul II (1464–1471). Nicolaus's earliest copies of the *Geography* incorporated a world map based on Ptolemy's first projection, but by the mid-1460s he had adopted the technically more demanding second projection. The northwest part of Nicolaus's map diverges from Ptolemy in adding Sweden, Norway, and Greenland. See Fischer 1932a, 1:352–356. © Biblioteca Apostolica Vaticana.

*Plate 3.* The fourth regional map of Libyē in Ptolemy's cylindrical projection, from Urb. lat. 277, ff. 99v–100r. As drawn in the manuscripts of the A version, this map portrays the entire continent of Libyē (Africa). This sumptuous copy of the *Geography* in the Latin version of Jacopo d'Angelo was completed in 1472 by the Florentine workshop of Vespasiano da Bisticci; the text was copied by Hugo Comminelli. See Fischer 1932a 1:365–375. © Biblioteca Apostolica Vaticana.

*Plate 4.* The third regional map of Europe in Ptolemy's cylindrical projection, from Urb. lat. 277, ff. 78v–79r (cf. Plate 3). Ptolemy's caption (8.4) is on the page facing the map. © Biblioteca Apostolica Vaticana.

*Plate 5.* Map of Aithiopia Below Egypt from the Greek manuscript B. L. Burney 111, f. 61r (c. 1400; see p. 45). This map is typical of the B version in its comparatively crude execution. The region portrayed includes the upper Nile and its sources. By permission of The British Library.

*Plate 6.* Map of the world in Ptolemy's second projection, from the 1482 Latin edition printed by Lienart Holle at Ulm. This was the first edition of the *Geography* to have woodcut maps. The maps derive from one of Nicolaus Germanus' later copies of the *Geography* (now at Schloss Wolfegg, Wurtemberg); the world map is inscribed with the name of the engraver, Johannes of Armsheim. Most copies of this edition have the maps colored, as in the present instance. Department of Printing and Graphic Arts, The Houghton Library, Harvard University.

*Plate 7.* The first regional map of Europe in the Donis projection, from the Latin edition of Ptolemy printed by Pietro de la Torre (Petrus de Turre) at Rome in 1490. This was a reprinting of the rare 1478 edition printed at Rome by Conrad Sweynheym and Arnold Buckinck, using the same copper plates. The plates, which are of a very high quality of craftsmanship, were based on one of Nicolaus Germanus' copies, probably the Codex Ebnerianus dedicated by Nicolaus to Duke Borso d'Este of Modena, which is now in the New York Public Library. Brown University Library.

*Maps 1–2* portray the continental outlines and principal features of Ptolemy's world map in his first and second projections. The data on which these maps, as well as Maps 3–7, are based are mostly derived from Nobbe's edition. These reconstructions may be compared with Plates 1, 2, and 6.

*Map 3* is a reconstruction of Ptolemy's second regional map of Europe, based on Cuntz's text of 2.7–2.10 (cf. Plate 4). We have omitted the names of peoples, the placements of which are only roughly indicated by Ptolemy. To make a logically consistent map we have diverged from the text in the following details:

1. The last point given for the boundary of Gallia Aquitanica is supposed to lie on the Pyrenees. Cuntz's point is about ½° too far north. We have moved it south to meet the mountains.
2. Avaricum in Aquitanica lies slightly to the east of the river Liger if that river is drawn rectilinearly, which would put Avaricum in Lugdunensis. This is probably not an error in the text, but just that Ptolemy did not note a slight bend in the Liger. We have given the river a bend.
3. As noted in our translation, the coordinates of a "midpoint" of the Sequana (in Gallia Lugdunensis) have probably dropped out of the text. One sign that this has happened is that the point identified as the midpoint is actually close to the source, and no source is indicated. Moreover, a straight line joining the mouth to this "midpoint" leaves Iuliobona, Condevincum, Ingena, and Lucotecia all on the wrong side of the river; whereas Ratomagus in Beltica has to be kept on the east side of the

river. We have made the river to keep on the appropriate side of all the cities by adding curves.

4. The border of Lugdunensis with Narbonensis is supposed to be along the Cemmena Mountains. Cuntz's "midpoint" doesn't make sense if the range goes all along the border, as Ptolemy seems to want. We have guessed at a plausible arc for the range. The border cannot follow the range's line exactly, because Lugdunum and Cabullinum (and the Rhodanus itself) have to stay on the south of the range, but must be within Lugdunensis.
5. Ptolemy has left the curvature of the Arar and Dubis (Narbonensis) near their sources to our discretion. A curve must be given to keep them from intersecting the Rhodanus.
6. We have given the Rhenus (Beltica) north of the Obrinca a bend sufficient to put all cities on the west side.

*Maps 4–8* portray large regions of the *oikoumenē*, with most localities marked that are mentioned in the chapters of the *Geography* translated in this book. For each map, version *a* is drawn according to Ptolemy's coordinates, using the cylindrical projection prescribed by Ptolemy in 8.1 for regional maps. Version *b* gives the known or presumed locations of most of the same places according to modern geographical knowledge. In these maps we mark with a dot physical features that Ptolemy treats as well-defined points, as well as towns and cities. See the Geographical Index for further discussion of the individual localities.

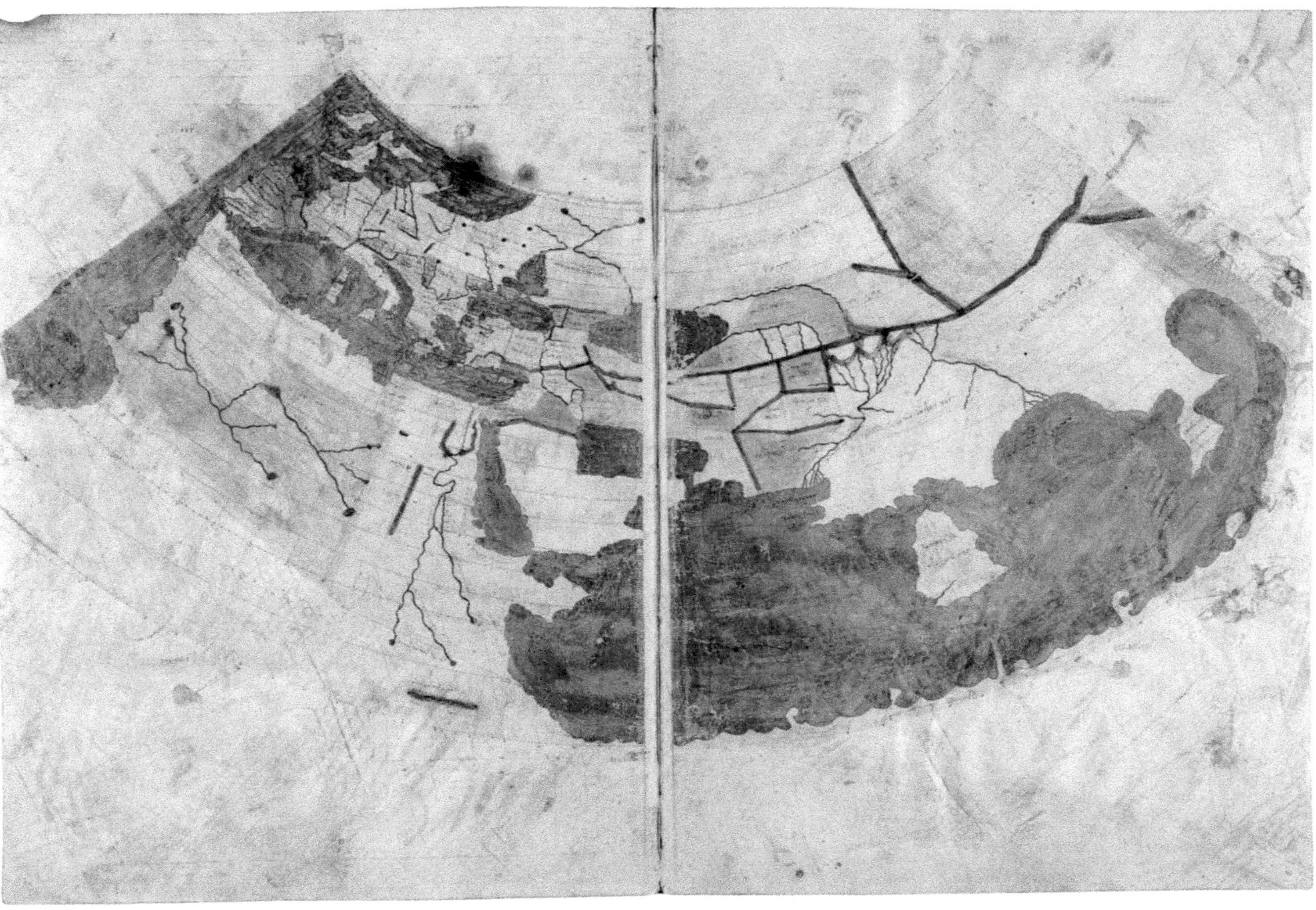

PLATE 1. Map of the world in Ptolemy's first projection (Urb. gr. 82, ff. 60v–61r)

PLATE 2. Map of the world in Ptolemy's second projection (Urb. lat. 274, ff. 74v–75r)

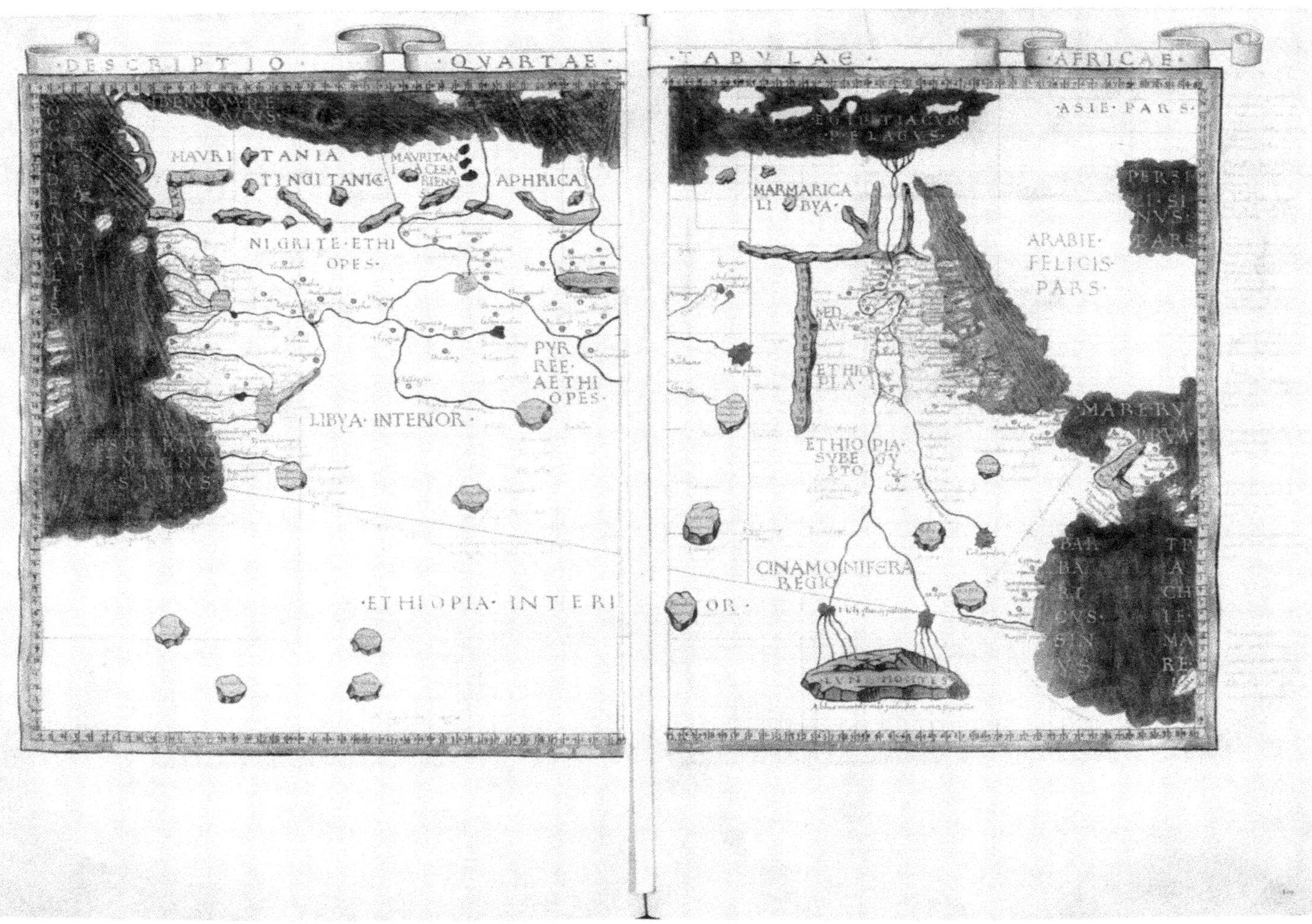


PLATE 3. The fourth regional map of Libyē in Ptolemy's cylindrical projection (Urb. lat. 277, ff. 99v–100r)

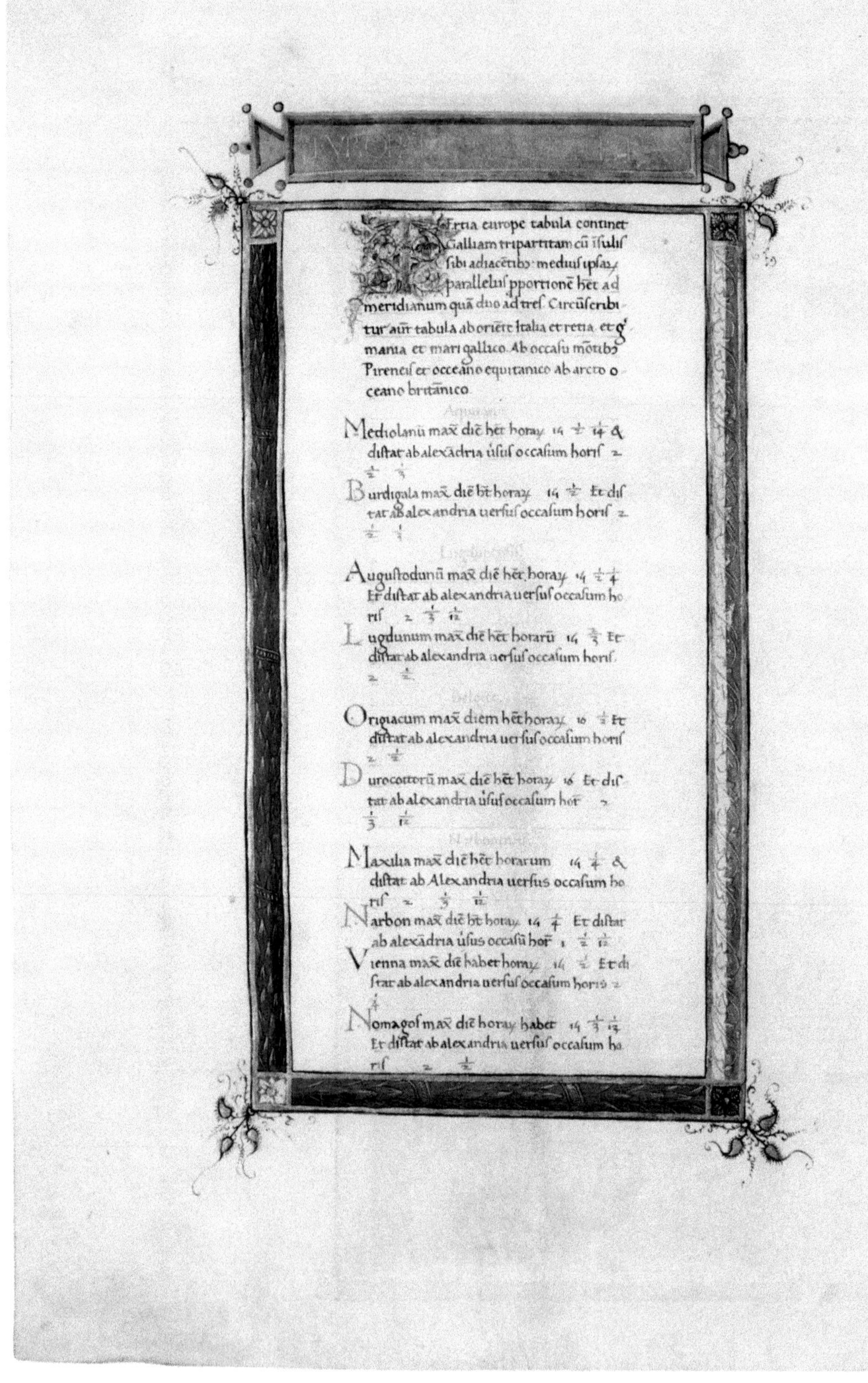

Tertia europe tabula continet Galliam tripartitam cū īsulis sibi adiacētibꝫ: medius ipsaꝝ parallelus pportionē hēt ad meridianum quā duo ad tres. Circūscribitur aūt tabula ab oriēte Italia et retia et germania et mari gallico. Ab occasu mōtibꝫ Pireneis et occeano equitanico ab arcto oceano britānico

Mediolanū max dīe hēt horaꝝ 14 1/2 1/4 & distat ab alexādria ūsus occasum horis 2 1/2 1/3

Burdigala max dīe hēt horaꝝ 14 1/2 Et distat ab alexandria uersus occasum horis 2 1/2 1/4

Augustodunū max dīe hēt horaꝝ 14 1/2 1/4 Et distat ab alexandria uersus occasum horis 2 1/3 1/12

Lugdunum max dīe hēt horarū 14 2/3 Et distat ab alexandria uersus occasum horis 2 1/2

Origiacum max diem hēt horaꝝ 16 1/2 Et distat ab alexandria uersus occasum horis 2 1/2

Durocottorū max dīe hēt horaꝝ 16 Et distat ab alexandria ūsus occasum hor 2 1/3 1/12

Maxilia max dīe hēt horarum 14 1/4 & distat ab Alexandria uersus occasum horis 2 1/3 1/12

Narbon max dīe hēt horaꝝ 14 1/4 Et distat ab alexādria ūsus occasū hor 1 1/2 1/12

Vienna max dīe habet horaꝝ 14 1/2 Et distat ab alexandria uersus occasum horis 2 1/4

Nomagos max dīe horaꝝ habet 14 1/3 1/12 Et distat ab alexandria uersus occasum horis 2 1/2

PLATE 4. The third regional map of Europe in Ptolemy's cylindrical projection (Urb. lat. 277, ff. 78v–79r)

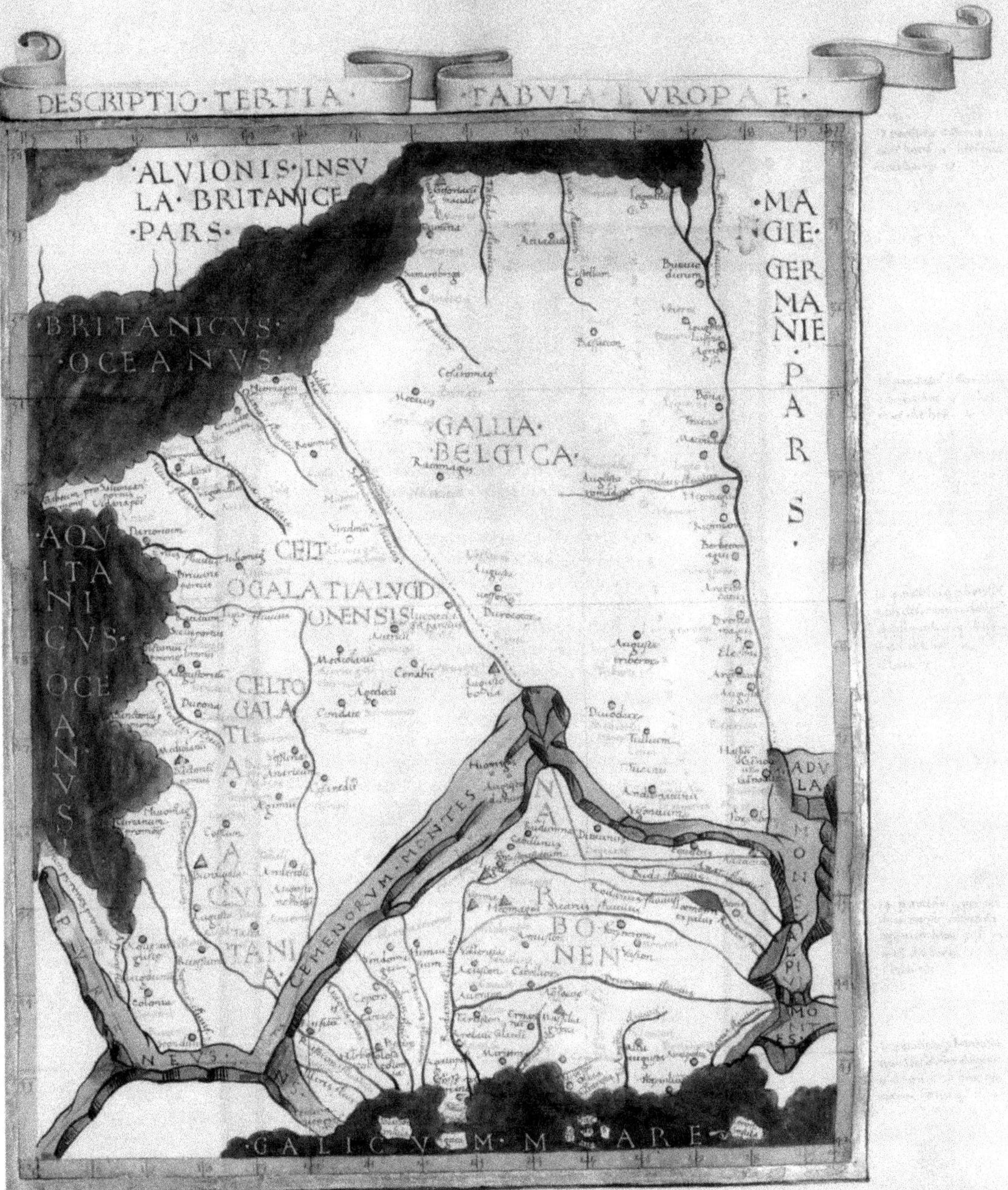
DESCRIPTIO·TERTIA·
·TABVLA·EVROPAE·
·ALVIONIS·INSVLA·BRITANICE·PARS·
·MAGIE·GERMANIE·PARS·
·BRITANICVS·OCEANVS·
GALLIA·BELGICA·
AQVITANICVS·OCEANVS·
CELTOGALATIA LVGDONENSIS
CELTOGALATIA AQVITANIA
CEMENORVM MONTES
NARBONENSIS
ADVLA
·GALICVM·MARE·

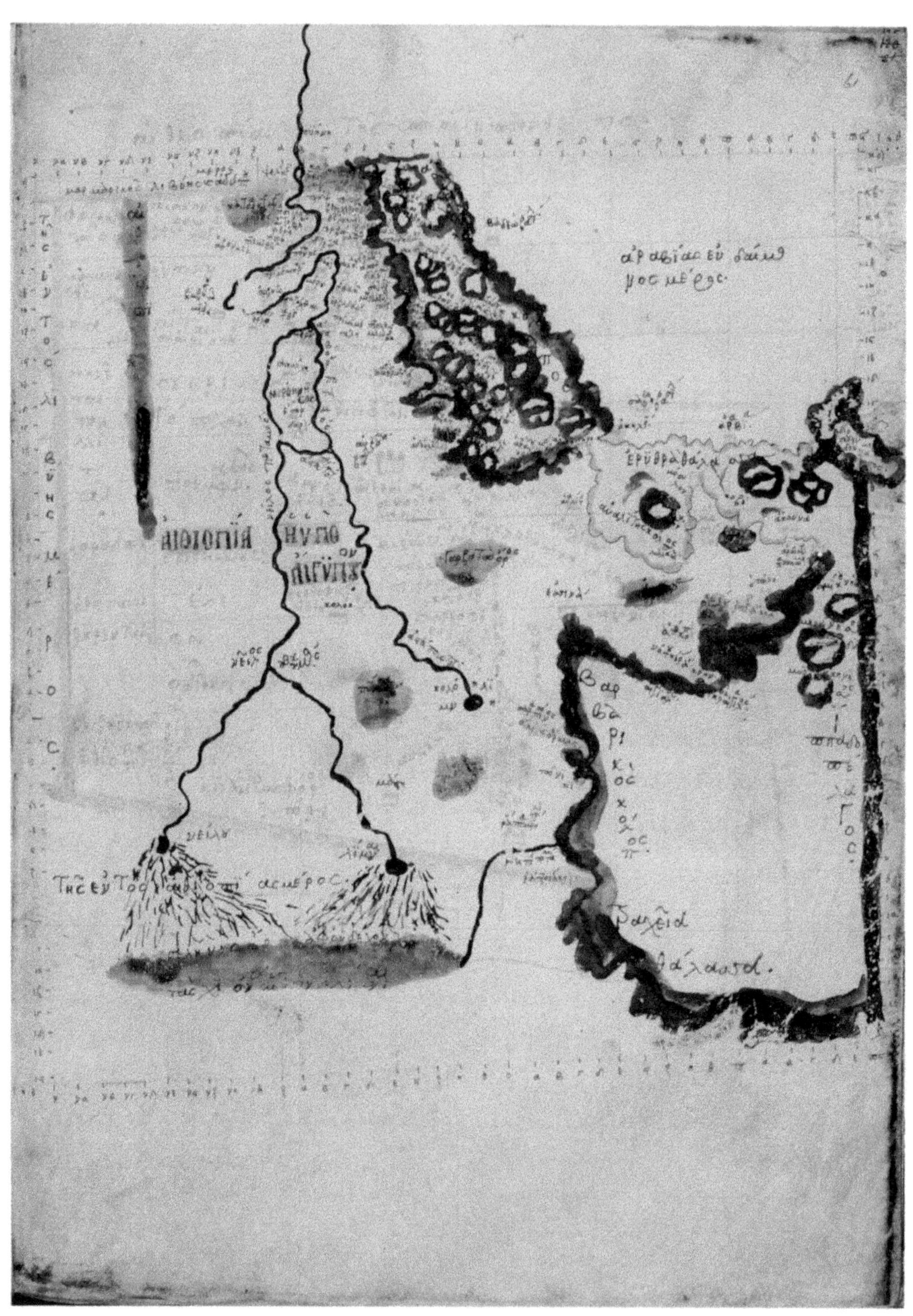

PLATE 5. Map of Aithiopia below Egypt (B. L. Burney 111, f. 61r)

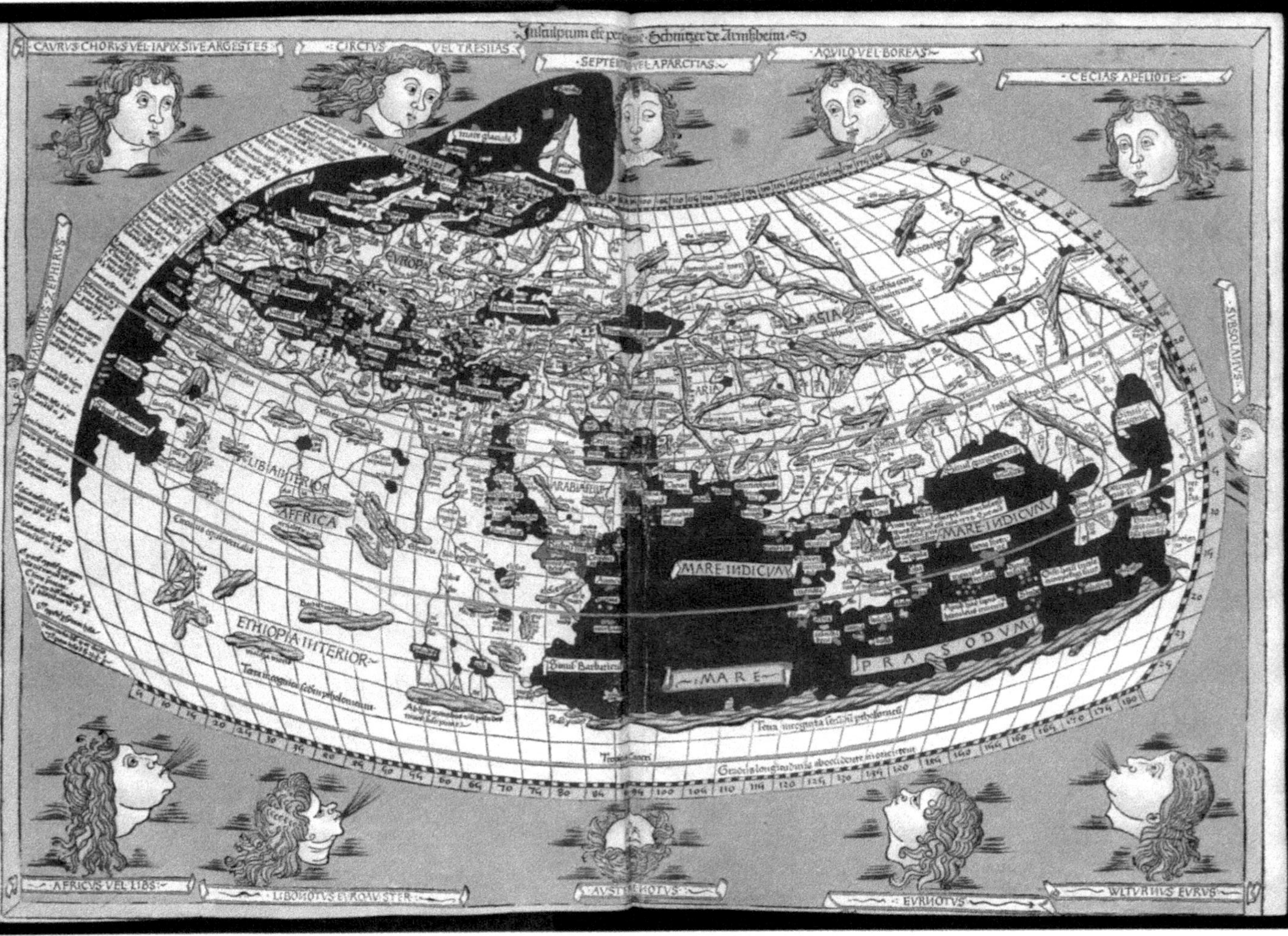


PLATE 6. Map of the world in Ptolemy's second projection (Ulm edition of 1482)

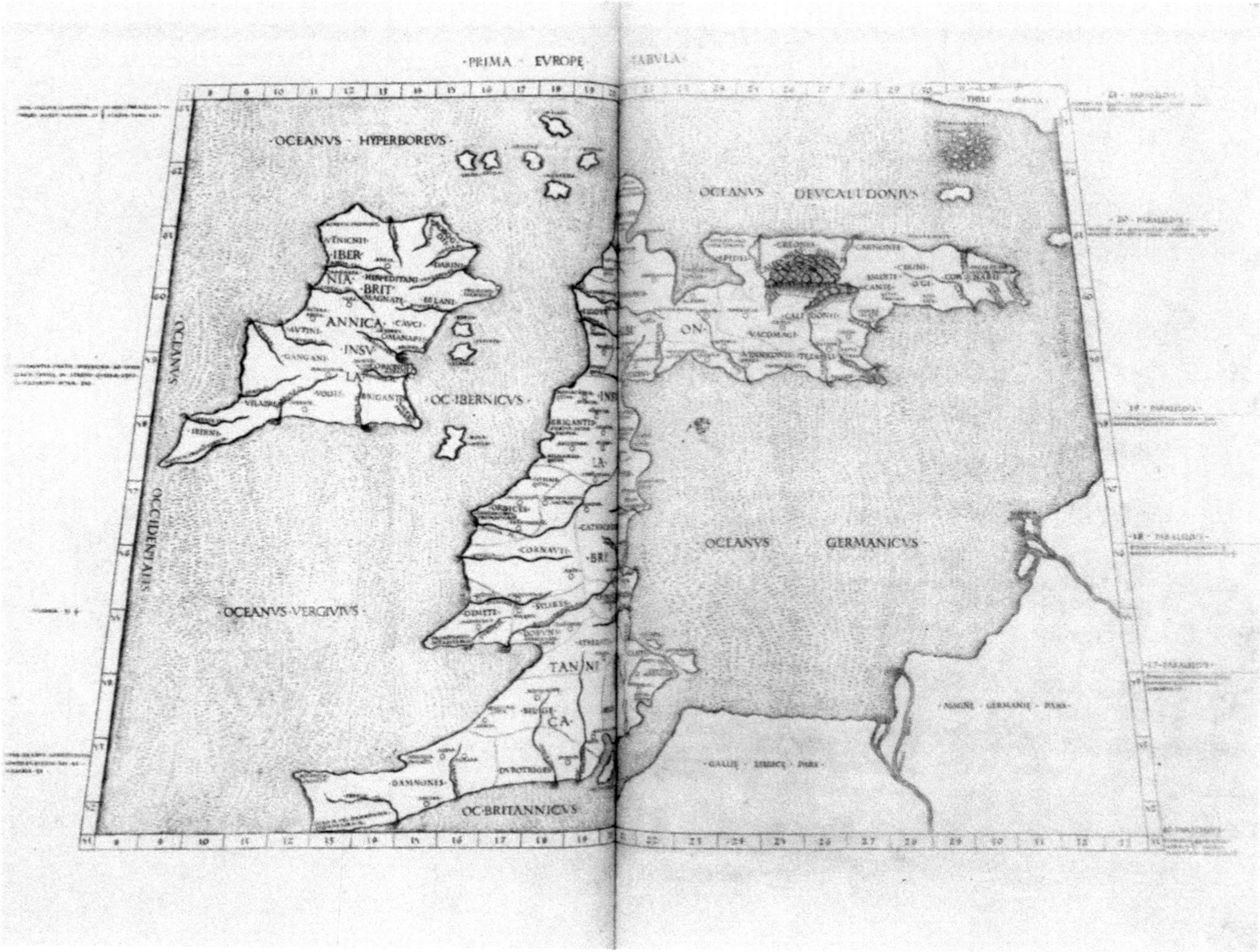

PLATE 7. The first regional map of Europe in the Donis projection (Rome edition of 1490)

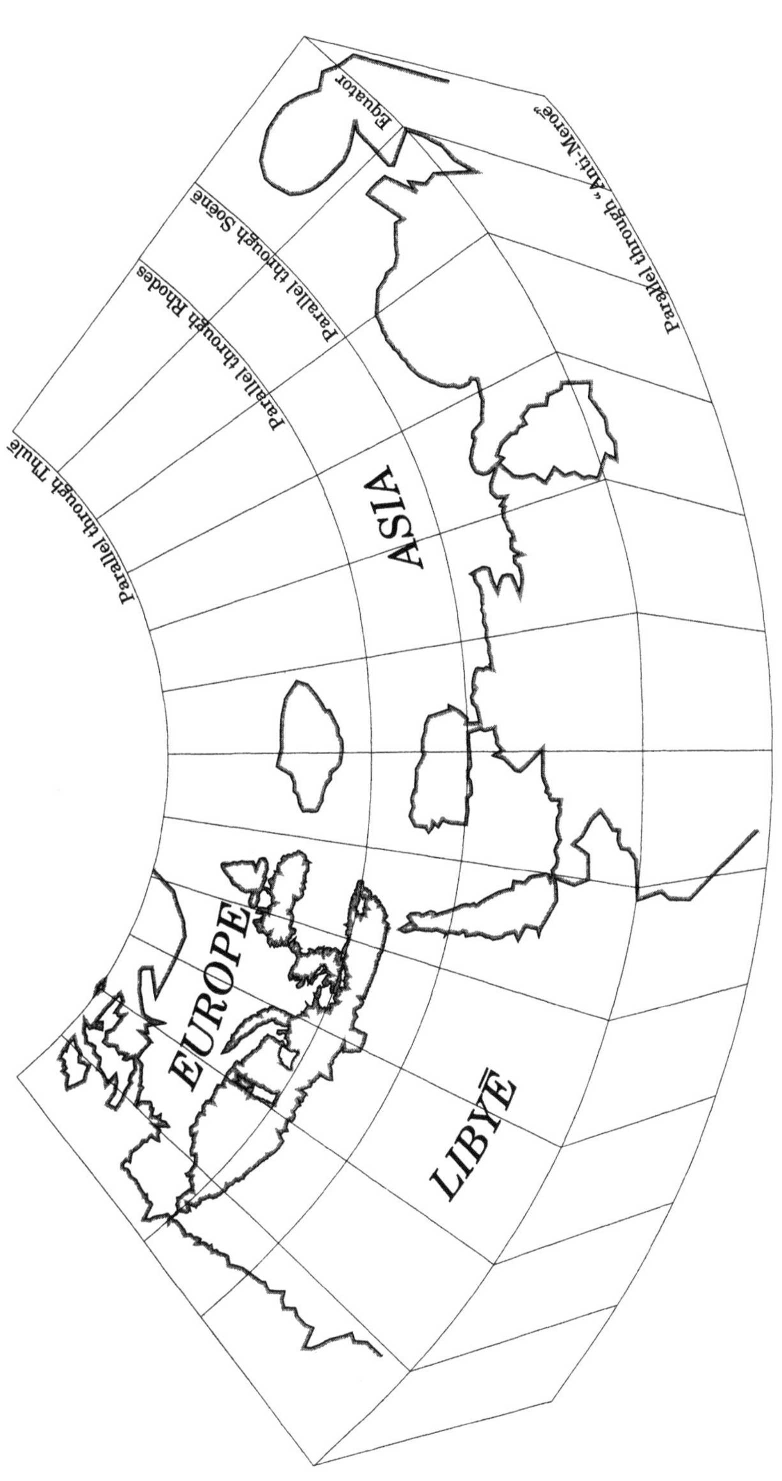


MAP 1. The *oikoumenē* in Ptolemy's first projection

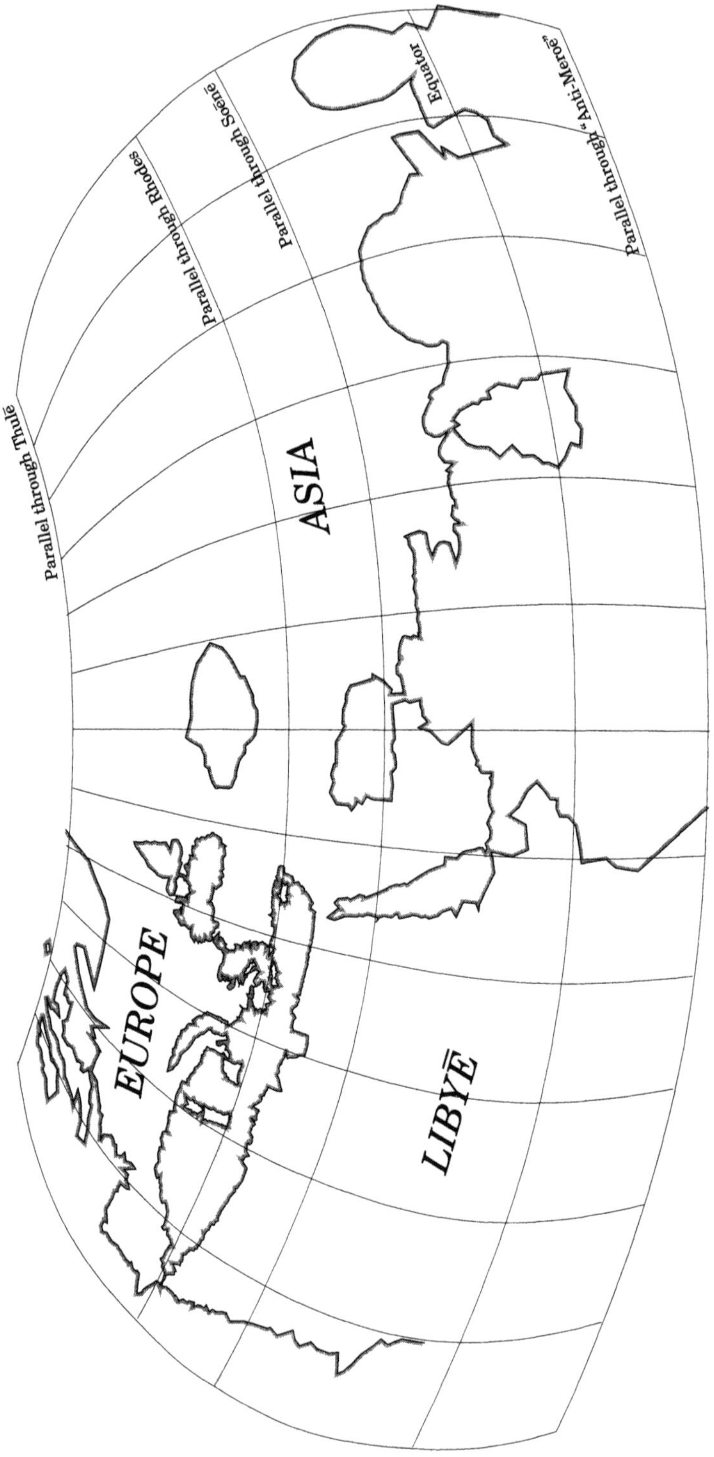

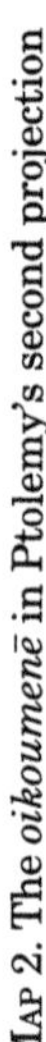

MAP 2. The *oikoumenē* in Ptolemy's second projection

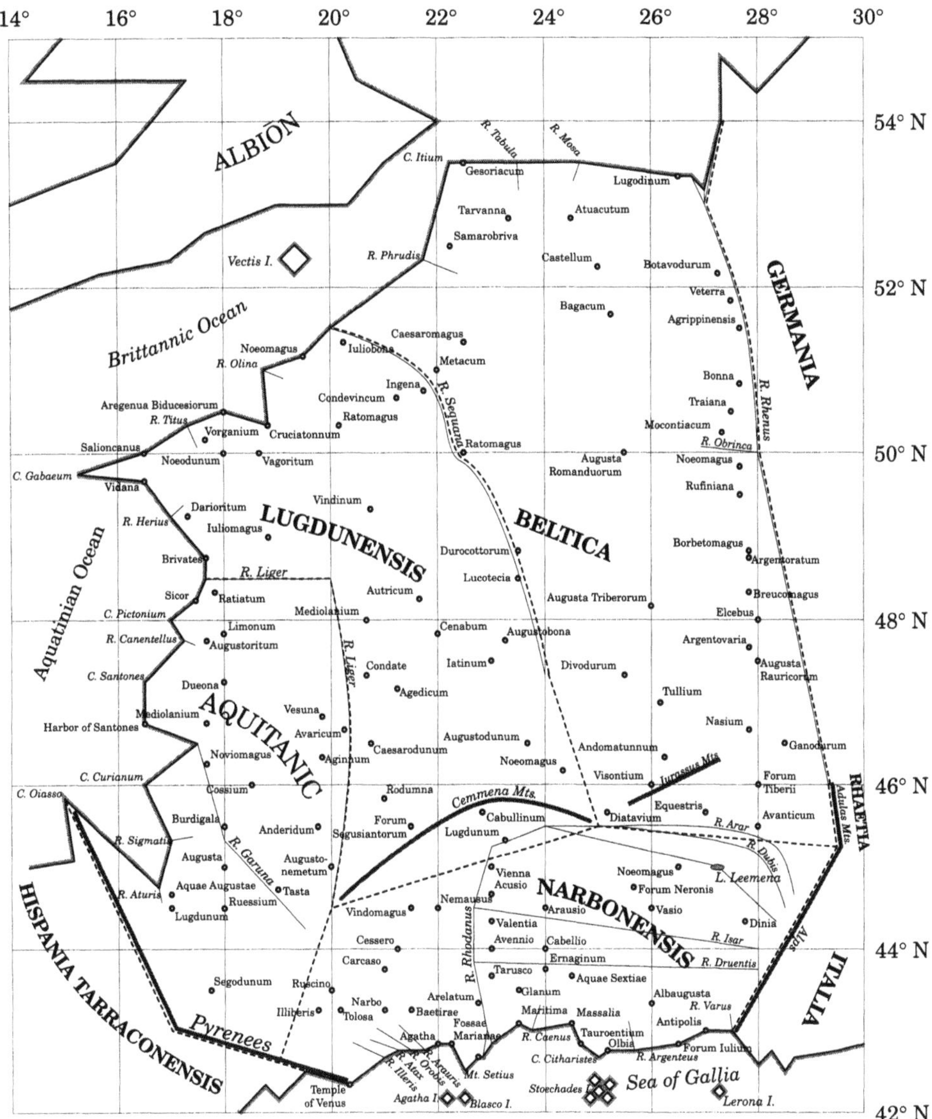


MAP 3. Ptolemy's third map of Europe (Gaul)

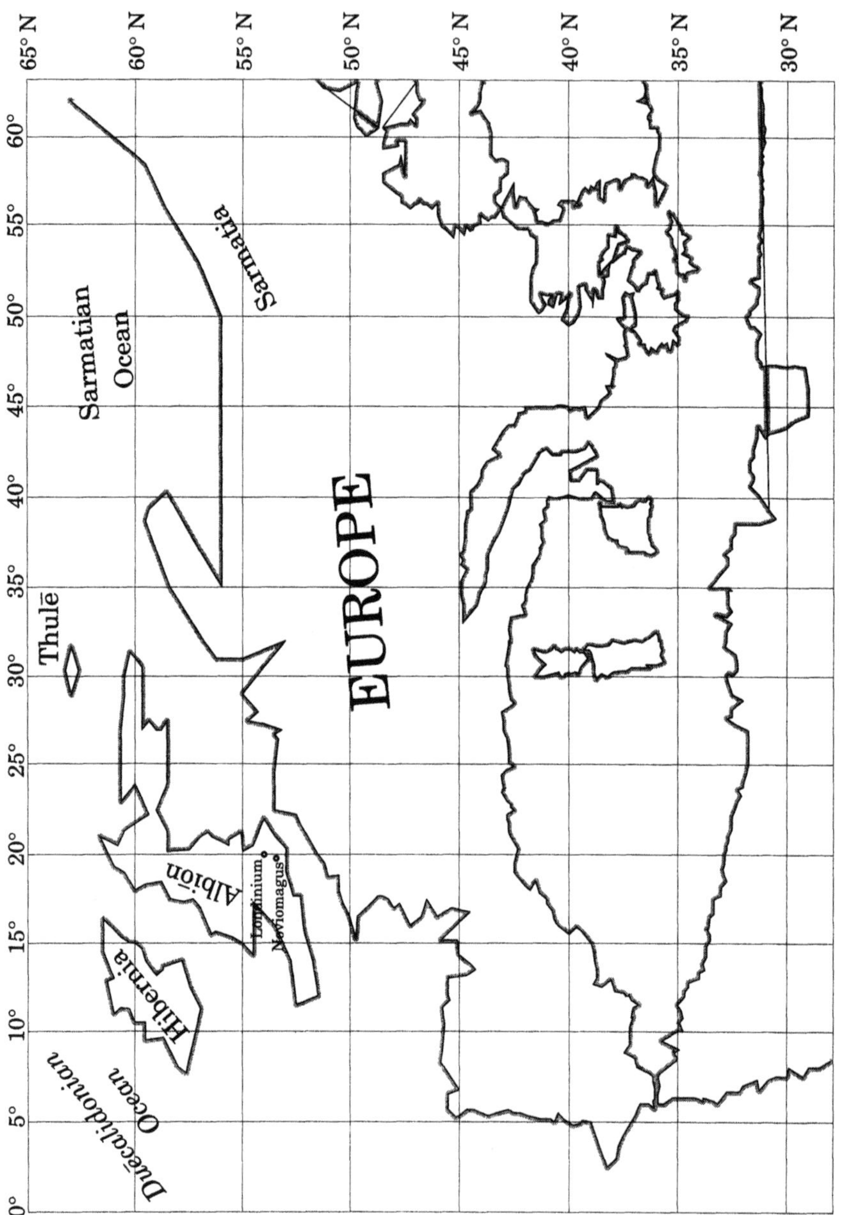


MAP 4A. Western Europe according to Ptolemy

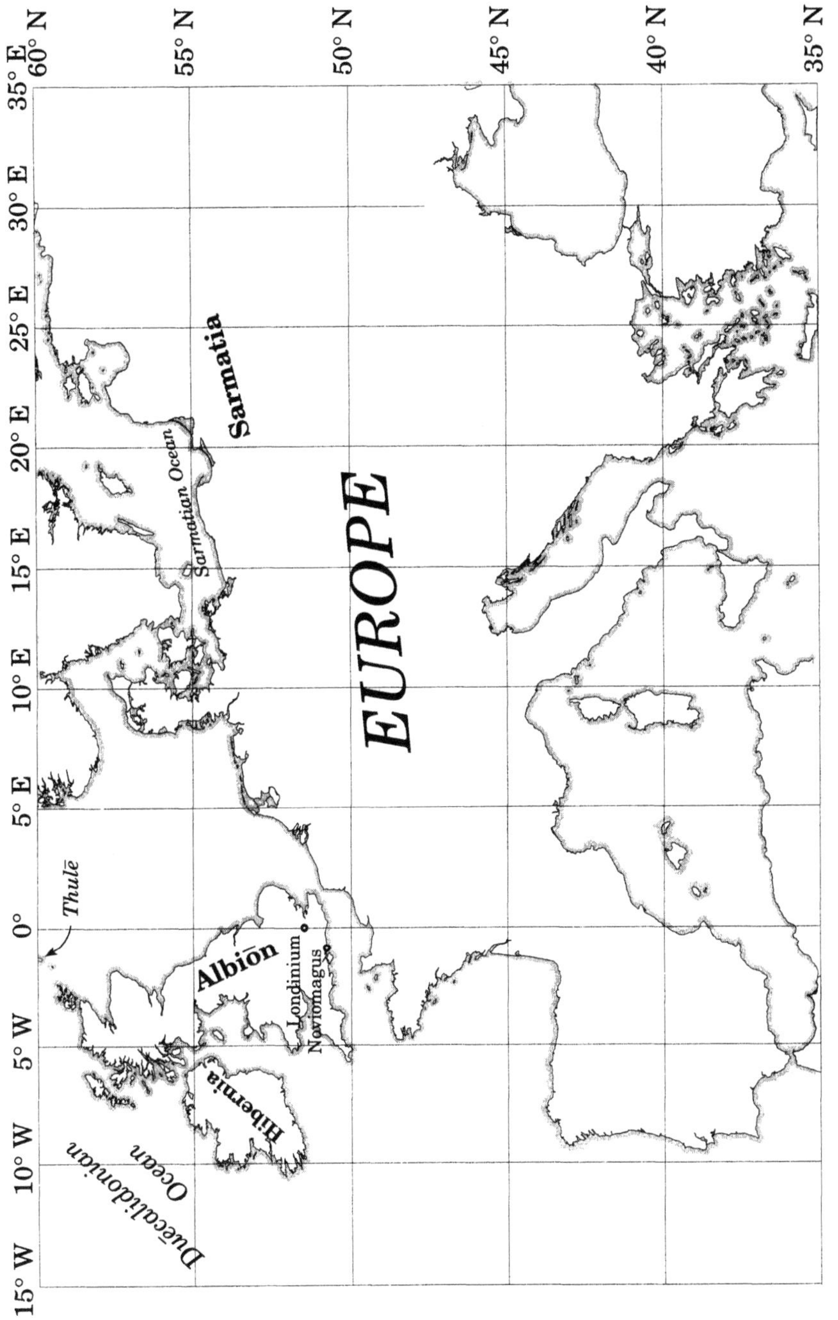

MAP 4B. Western Europe

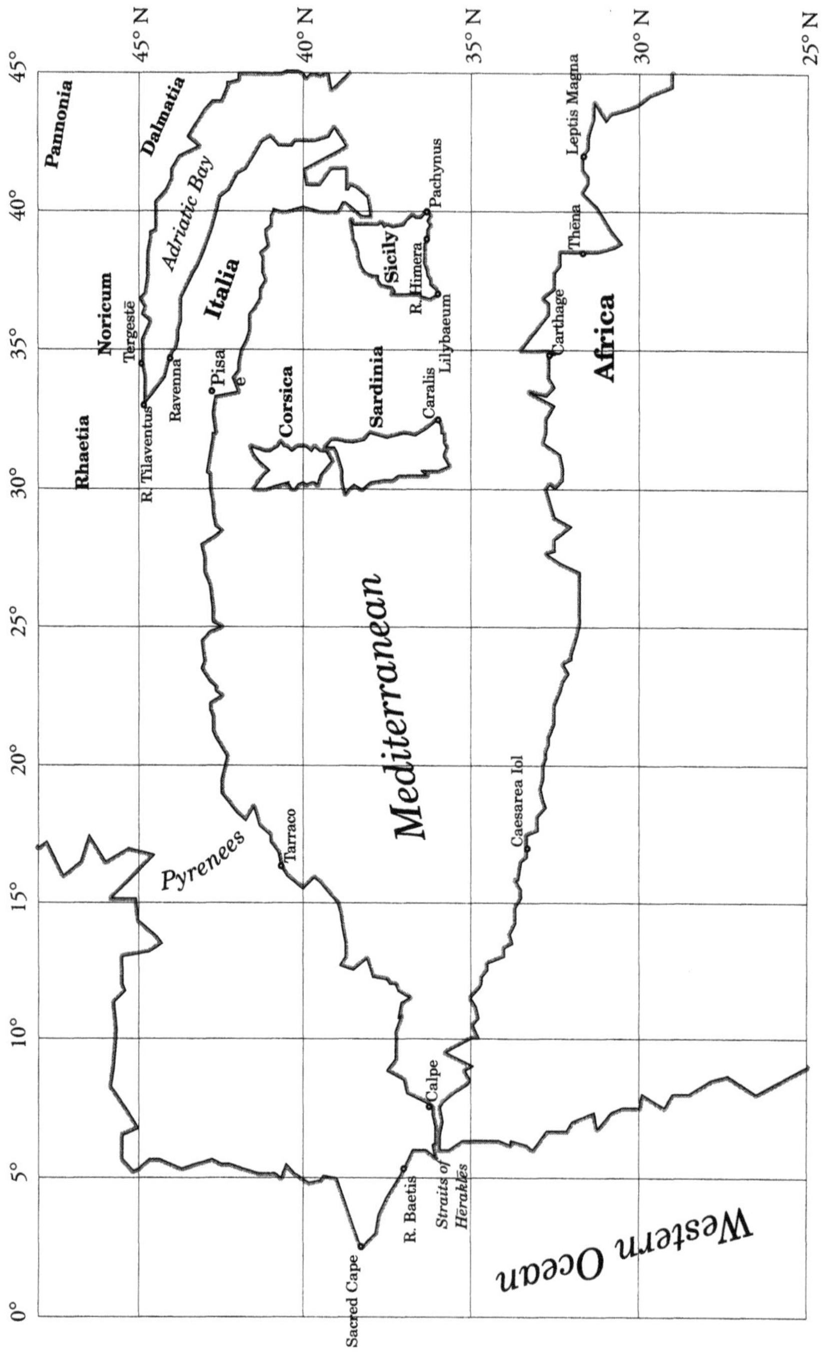

MAP 5A. The western Mediterranean according to Ptolemy

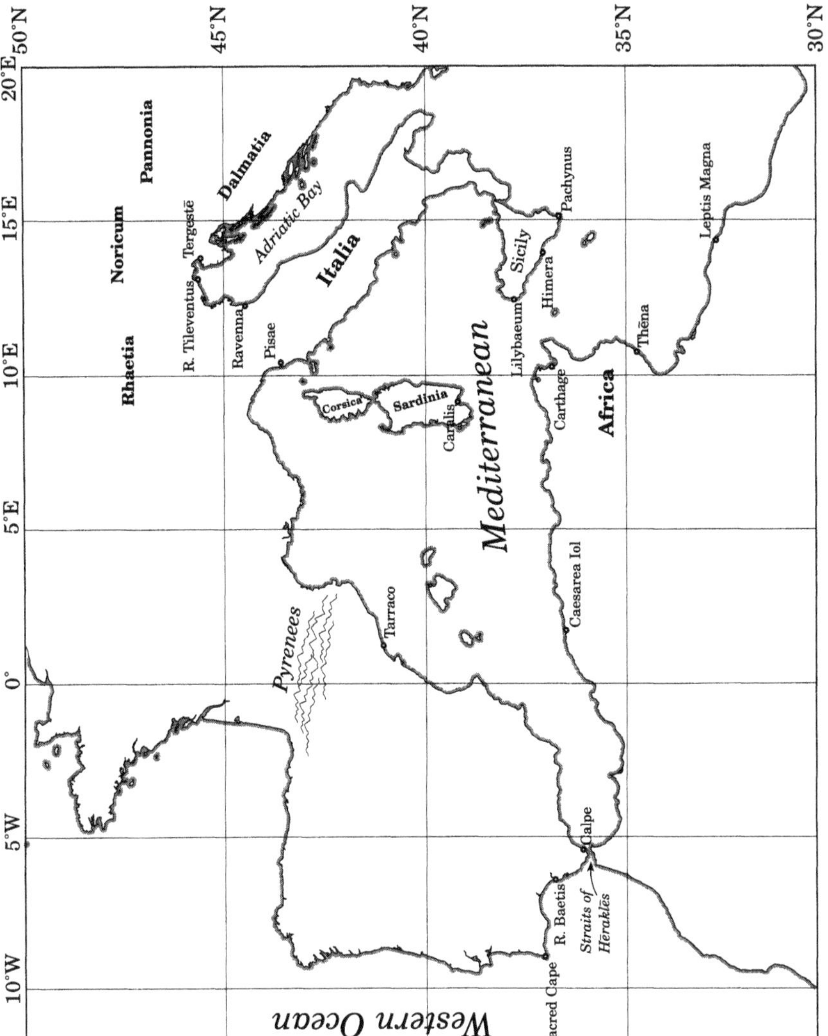


MAP 5B. The western Mediterranean

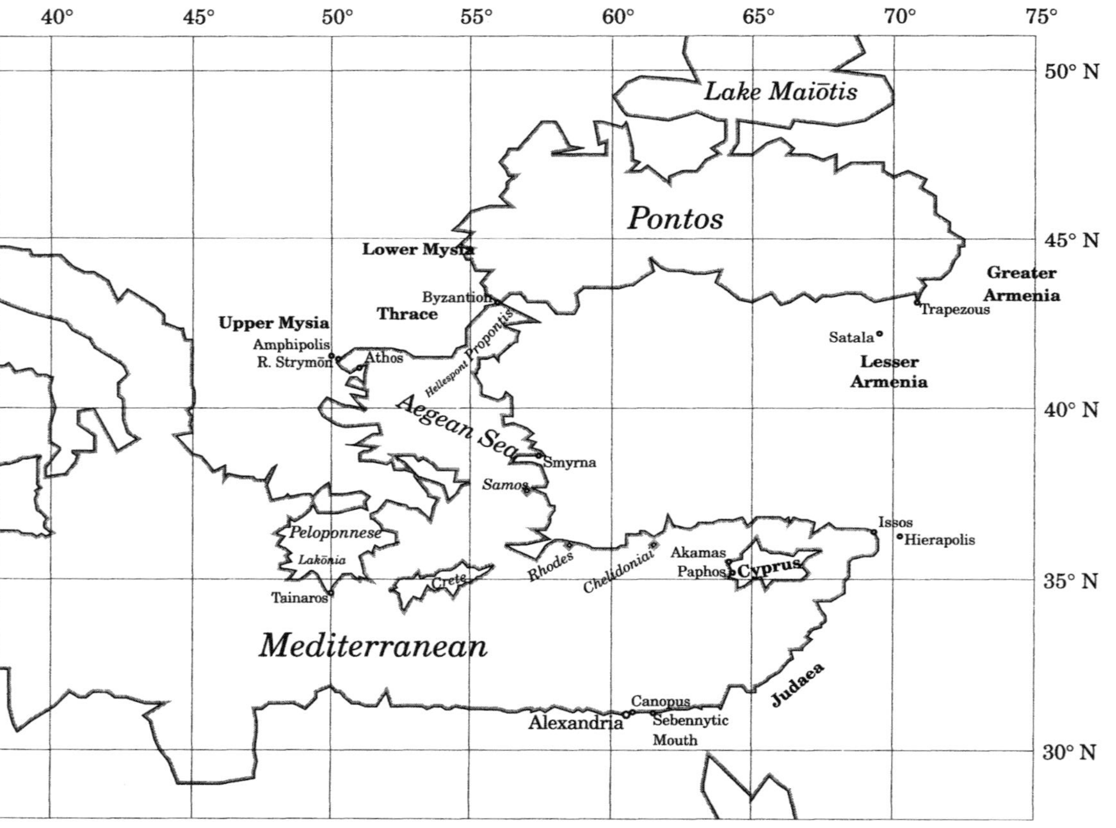


MAP 6A. The eastern Mediterranean according to Ptolemy

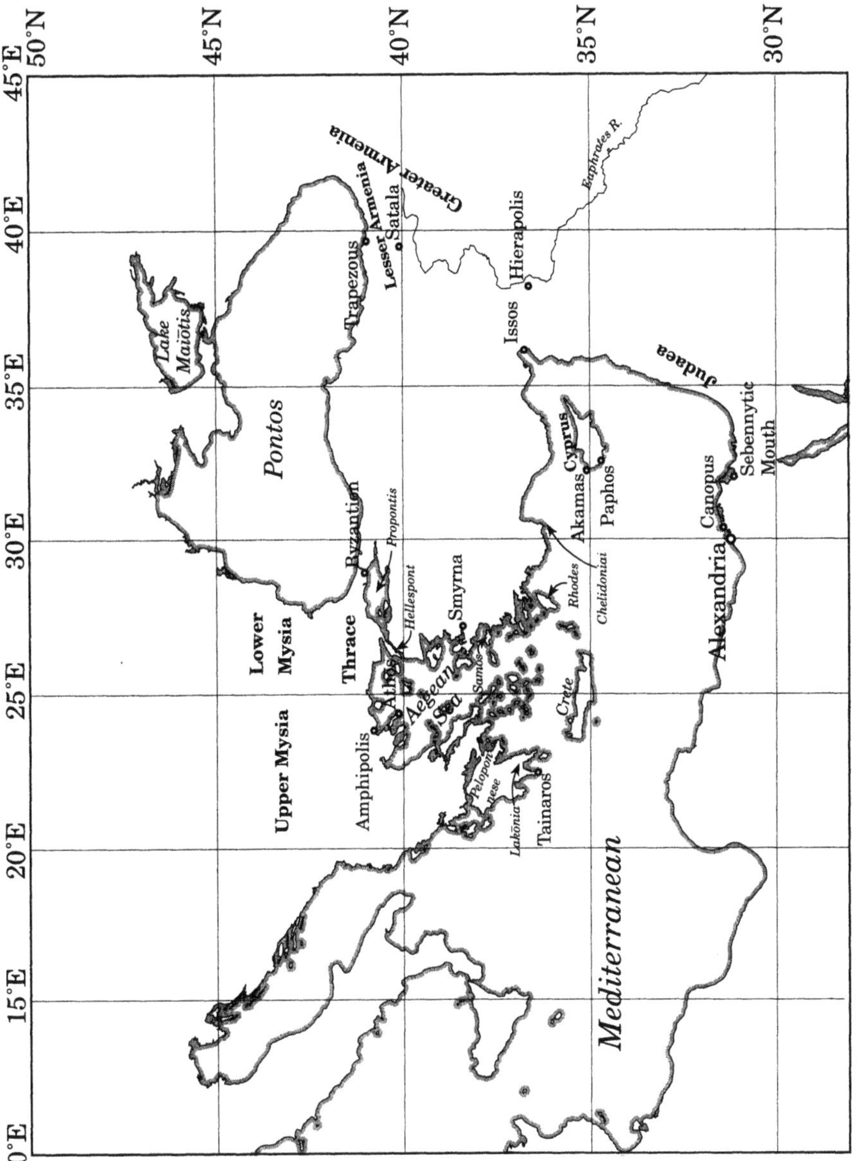


Map 6b. The eastern Mediterranean

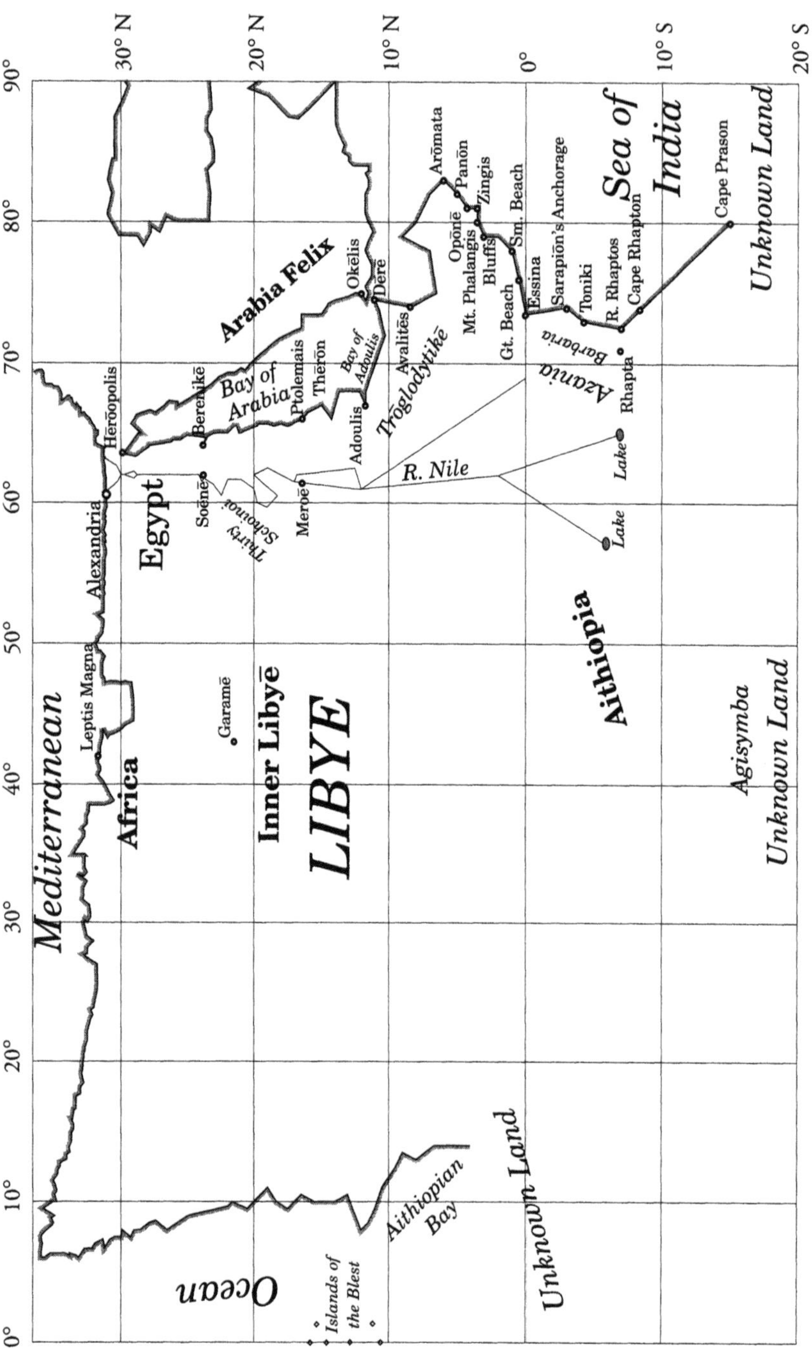


MAP 7A. Libyē (Africa) according to Ptolemy

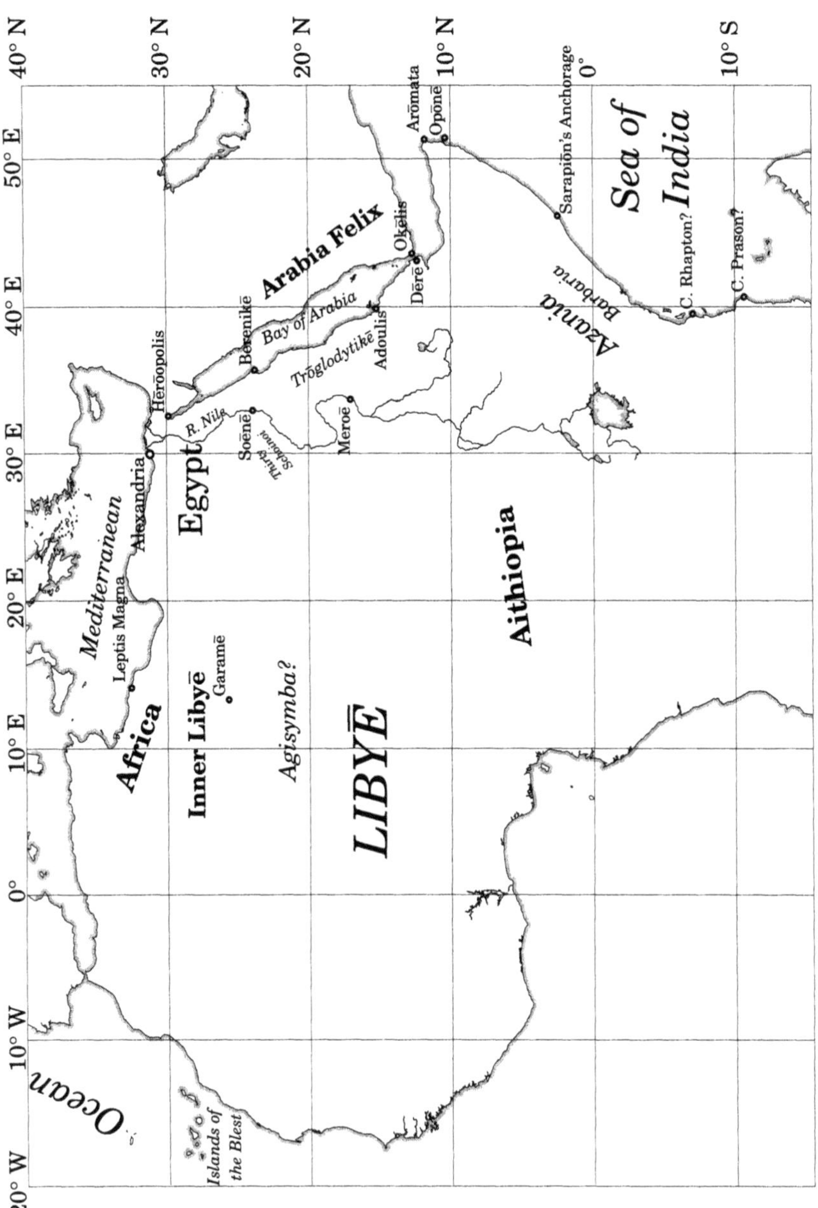


MAP 7B. Libyē (Africa)

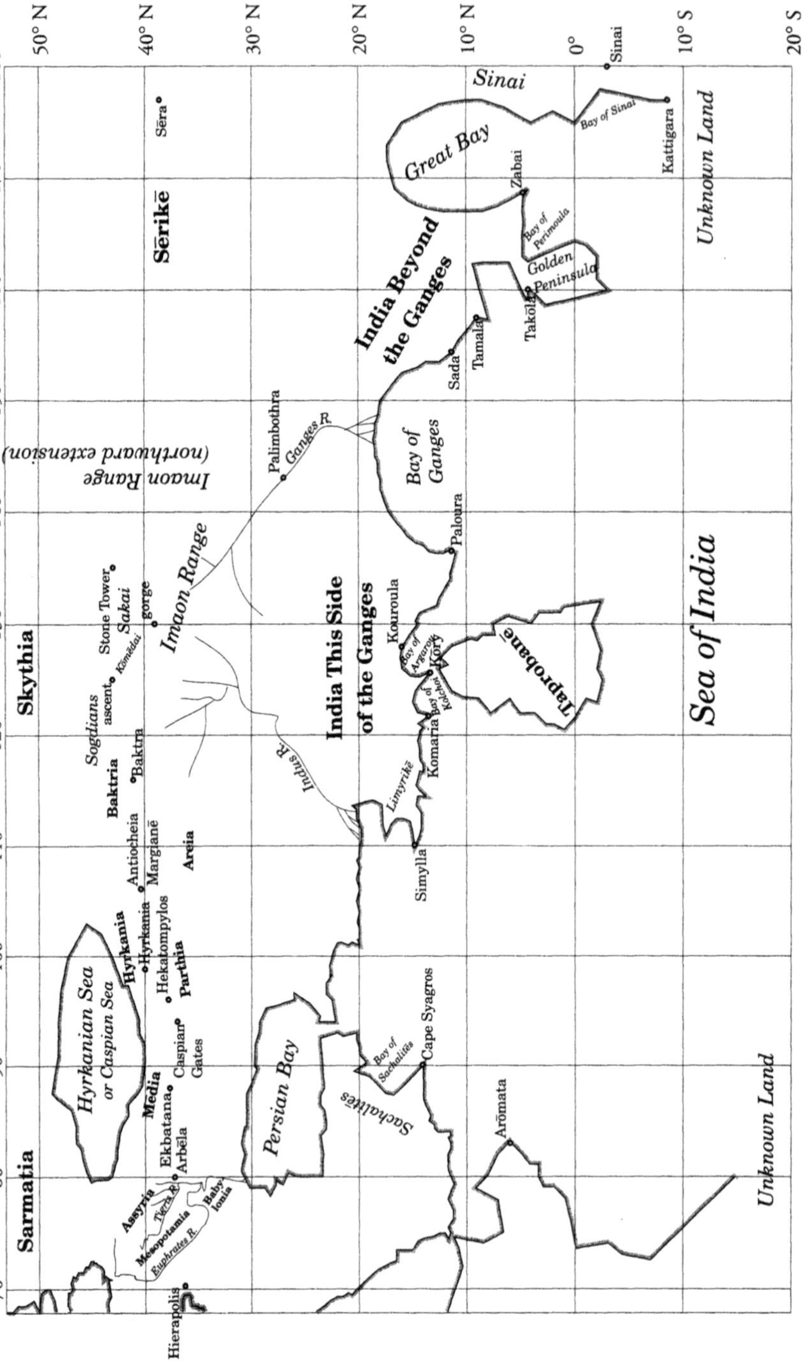

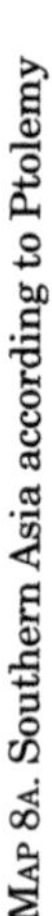


MAP 8A. Southern Asia according to Ptolemy

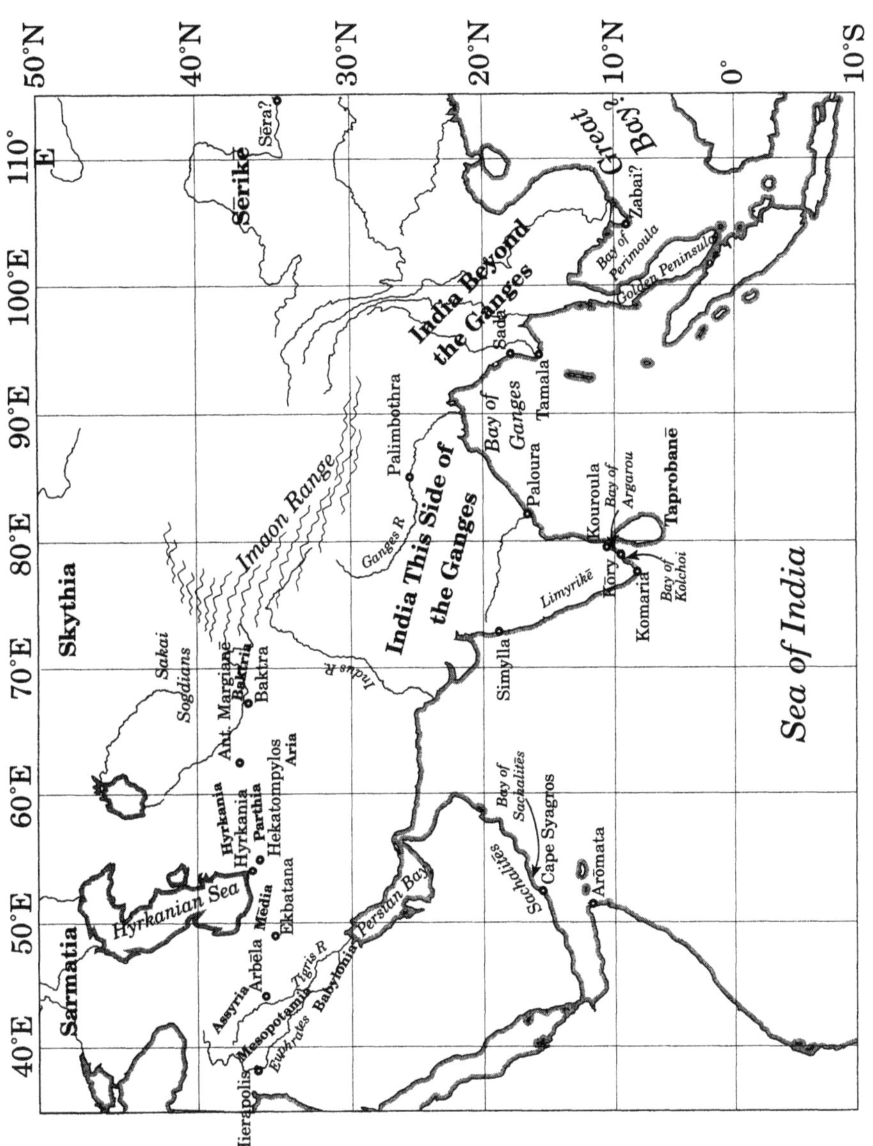


MAP 8B. Southern Asia

# APPENDICES

# *Appendix A*
# *The Journeys of Flaccus and Maternus to the Aithiopians*

Marinos based his calculation of the latitude of the southernmost known point of the interior of the African continent on two reports of Roman explorations. One was a military expedition led by a certain Septimius Flaccus, who, according to Ptolemy's first mention of the journey (1.8), set out "from Libyē," and who proceeded south for three months from the district ("people") of Garamē until he reached the Aithiopians. The second traveler, Julius Maternus, is said to have set out from the city of Leptis (or Lepcis) Magna, on the Mediterranean coast of the Roman province of Africa, and then to have accompanied the king of Garamē on a military campaign south from Garamē to an Aithiopian country called Agisymba. The duration of the journey from Garamē to Agisymba is stated in 1.8 as four months, although later (1.11) Ptolemy refers to it as taking four months and fourteen days. Ptolemy adds that Agisymba was subject to the king of Garamē, and that rhinoceros were to be found there. In a later passage (1.10), Ptolemy writes that both Flaccus and Maternus determined the distance from Leptis Magna to Garamē as 5,400 stades, that the journey from Leptis to Garamē took thirty days and the return journey twenty days, and that specific numbers of stades were reported for each day's progress from watering place to watering place on this desert route. It would appear, therefore, that notwithstanding the odd reference to Flaccus' coming from Libyē, both expeditions set out from Leptis.[1]

Desanges (1978, 197–213) offers some interesting and plausible conjectures about the expeditions. Since Pliny the Elder makes no mention of them in his full treatment of Roman activity in this part of northern Africa (*Nat. Hist.* 5.34–38), they would seem to have come after his death in A.D. 79. The references to rhinoceros in connection with Maternus' journey to Agisymba could be connected with evidence from coins and literary sources that rhinoceros were produced at

[1]"Libyē" normally means the African continent for Ptolemy, though there was also a district by this name west of Alexandria, and Ptolemy calls the Sahara "desert Libyē." The simplest explanation of the word here, however, is that it was used by Marinos as a Greek name for the Roman province of Africa.

Rome during the reign of Domitian, apparently between the years 83 and 92. If the purpose of Maternus' expedition was in fact to procure exotic animals for the Roman spectacles, then Agisymba should have been at the northern limit of the region where rhinoceros were to be found in the first century; and although this limit cannot be established with certainty, it may have been much further north than in modern times. The implication of Ptolemy's text that the king of Garamē regarded the people of Agisymba as his subjects also points to a location in the Sahara. Flaccus' military expedition looks as if it might have been a pacification of the people of Garamē—there had been earlier Roman campaigns against them about 20 B.C. and A.D. 78—and might be situated between the latter date and Maternus' expedition.[2]

Marinos initially calculated from the reports of Flaccus and Maternus that Agisymba lay 24,680 stades south of the equator (1.8), a distance that he afterward arbitrarily reduced to just 12,000 stades. Ptolemy does not tell us how Marinos arrived at the first figure, but it is possible to guess at his line of reasoning from the information that Ptolemy does provide. First, in 1.10 Ptolemy is evidently objecting to Marinos' estimate of the distance of Garamē from Leptis. He does not give Marinos' result, but he implies that Flaccus and Maternus had stated this distance as 5,400 stades based on a journey of twenty days, whereas Marinos preferred to use thirty days for the length of the journey. This means that Flaccus and Maternus assumed an (average) rate of travel of 270 stades (approximately 50 kilometers) per day, which seems rather fast. Supposing that Marinos adopted this speed, and applied it to an interval of thirty days for the stage from Leptis to Garamē, and an interval of four months, say 120 days, from Garamē to Agisymba, always due south, then he would have situated Garamē 8,100 stades south of Leptis, and Agisymba 32,400 stades south of Garamē. We do not know the precise latitude that Marinos assigned to Leptis; but Ptolemy places it at 31⅔° (i.e., 15,833 stades) north of the equator, and Marinos' position is not likely to have been very different for an important city on the Mediterranean. Hence Marinos' Garamē might have been 7,733 stades (or 15½°) north of the equator, and his Agisymba would have been 24,666 stades south of the equator. The discrepancy with the figure that Ptolemy gives is so small that it appears very probable that Marinos worked in essentially this way. Aside from not using the additional fourteen days of travel mentioned in 1.11, it appears that Marinos used whatever interpretation of the data would give him the greatest possible latitudinal interval.

Ptolemy himself situates Garamē at longitude 43°, and latitude 21½° (4.6), so that it is 1° east and 10⅙° (5,083 stades) south of Leptis. Hence he has him-

[2]It is conceivable that Ptolemy's "Septimius Flaccus" is an error for Suellius Flaccus, who is known to have led a campaign in A.D. 86 against the Nasamones, a people inhabiting the district to the northeast of Garamē.

self not strictly followed the assertion of Flaccus and Maternus that Garamē was 5,400 stades due south of Leptis. In 4.9 Agisymba is described as an extensive region along the southern limit of his *oikoumenē*, which he established in 1.9–1.10 on ecological grounds.

# Appendix B
# Marinos' Calculation of the Latitude of Cape Prason

According to Ptolemy (1.8), Marinos calculated that Cape Prason on the east coast of Africa was 27,800 stades south of the equator (a figure that he subsequently reduced, as he also did for the latitude of Agisymba) on the basis of the number of days of sail from Ptolemais Thērōn on the Red Sea coast of Egypt to Cape Prason. Again, Ptolemy does not describe how Marinos arrived at his figure but restricts himself to a criticism of particular details. It would seem that the calculation was a sum of three components: the interval from Ptolemais to Arōmata at the extremity of the Horn of Africa, the interval from Arōmata to Cape Rhapton, and the interval from Cape Rhapton to Cape Prason.

Ptolemy does not directly discuss the first interval, and we might therefore attempt to reconstruct Marinos' value for its latitudinal extent from Ptolemy's map on the assumption that where he does not fault Marinos' methods, he is roughly following his data. In 4.7 Ptolemy places Ptolemais at longitude 66° and latitude 16⅚₁₂° (approximately 8,200 stades) north of the equator—which is precisely the parallel through Meroē, on which the longest day is 13 hours—and Arōmata at longitude 83°, latitude 6° north. The latitudinal component of the interval between the two places therefore amounts to approximately 5,200 stades on Ptolemy's map. But in 1.14, as part of a rather sophistical argument concerning the *longitudinal* extent of the known world, Ptolemy casually situates Arōmata at latitude 4¼° north (which is incidentally the parallel on which the longest day is 12¼ hours). The most plausible explanation of the inconsistency is that Ptolemy has here inadvertently given us Marinos' figure for the latitude of Arōmata. If Marinos put Arōmata 4¼° (2,125 stades) north of the equator, then he must have calculated the remaining latitudinal interval from Arōmata to Cape Prason as 29,925 stades.

For the next interval, Marinos relied on the reports of Diogenes and Theophilos (1.9), who were traders sailing respectively by the monsoon route from Arabia to the Malabar coast of India (Limyrikē) and by the route along the

African coast. What made their reports particularly useful to Marinos was that they had both been driven by storms directly from one end of the interval in question to the other, so that Marinos could hope to derive a more exact total for the time and distance involved than by adding up the many individual stages into which a trader would normally have broken up the journey. Diogenes was supposed to have sailed from Arōmata for twenty-five days before reaching the sources of the Nile, slightly north of Cape Rhapton. Theophilos' route was in the opposite direction, and covered more or less the same interval in twenty days. Marinos interpreted them as saying that the journey was in a straight north-south line, and he accepted Theophilos' estimate of the rate of sail as 1,000 stades per day.

Ptolemy tells us that the third interval, from Cape Rhapton to Cape Prason, was recorded by a certain Dioskoros as 5,000 stades. On Ptolemy's map, the two points are pretty nearly 5,000 stades apart, but Cape Prason is also placed east of Cape Rhapton, so that the latitudinal interval is only about 3,500 stades.

Since Marinos obtained almost exactly 30,000 stades for the whole latitudinal interval from Arōmata to Cape Prason, it seems that he must have applied Theophilos' estimate of 1,000 stades per day to Diogenes' twenty-five days of sailing, to get 25,000 stades as far as Cape Rhapton, and then added the whole 5,000 stades for the interval from Cape Rhapton to Cape Prason, i.e., assuming that the entire journey was due south. Again one gets the impression that Marinos was attempting to establish an extreme upper bound for the distance rather than the most plausible estimate.

In locating these places on his own map, Ptolemy began by assuming (for no very obvious reason) that Cape Prason would have been roughly as far south of the equator as Agisymba, the southernmost region of the interior of Africa reportedly reached by travelers from the Greco-Roman world. He seems to have tried to give as much space for the coast from Arōmata to Cape Rhapton as he dared by pushing Arōmata a little further north than Marinos had put it, and by deducting a little from the interval between Cape Rhapton and Cape Prason and inclining this stage enough to the southeast so that Cape Rhapton could lie right at latitude $8\frac{5}{12}°$ south, the parallel on which the longest day is $12\frac{1}{2}$ hours. Even so, he was left with only about 7,200 stades for the latitudinal interval between Arōmata and Cape Rhapton, which is less than a third of what Marinos found for it. As it turns out, Ptolemy hit upon a quite reasonable latitude for Cape Rhapton, since the most likely candidates for this place are between 5° and 8° south of the equator (Casson 1989, 141).

# *Appendix C*
# *The Trade Route across Central Asia*

The first of Marinos' two attempts to determine the longitudinal extent of Asia involved converting the stages of a route across the continent from Hierapolis in Syria to Sera, the chief city of the "Silk People" (Sēres) to a total distance in stades measured along the parallel 36° north of the equator (1.11–1.12). The journey in question was evidently one version of the "Silk Road," the route—or rather, the network of routes—by which goods traveled between the Mediterranean and China.

The starting point of the route, so far as it concerns Marinos' calculations, was "the crossing of the Euphrates at Hierapolis." (Hierapolis was actually to the west of the river, but there were places for crossing close by, including an important bridge at Zeugma.) Marinos was working from a list of stade distances along a trade route that in large part followed the so-called Royal Roads traversing the Parthian empire (cf. the list compiled by Isidoros of Charax, pp. 26–27), and apparently added up distances as if the journey continued more or less straight east as far as the Stone Tower, which marked the end of the part of the route for which stade distances were available. For much of this route, Marinos seems not to have had indications of the directions, since Ptolemy's examination of Marinos' data depends on a general sense of the latitudes of the countries through which the road passed.

Ptolemy is willing to admit that the first stages, from Hierapolis across the Euphrates and Tigris, through the old Mēdian capital Ekbatana and the Caspian Gates (a pass through the Elburz mountains), to the Parthian capital Hekatompylos, all lay along the parallel through Rhodes (36° north of the equator). In Isidore's version, however, the road tended southeast along the Euphrates between Hierapolis and the point where it crossed the Tigris (at Seleucia), and again northeast from there to the Caspian Gates. But since Ptolemy expressly states that Marinos' route traversed Assyria rather than Mesopotamia, it does look here as if Marinos' sources knew of a more direct road that cut directly east.[1]

[1]A map showing the Royal Road is in *The Cambridge History of Iran* 3.2 (Cambridge, 1983), 25.

After Hekatompylos, the Royal Road continued with a generally northeast trend beyond Hyrkania to Antiocheia Margianē, and then turned almost straight south through Areia, and then northeast again to Baktra. Ptolemy gives the sequence as Hyrkania, Areia, and then Antiocheia, which looks like an erroneous transposition. He is more or less correct, however, in stating that a direct route from Antiocheia to Baktra would head east.

Up to this point we have been dealing with a list of localities the positions of which are at least approximately known to us. The continuation, which describes the route across the great system of mountains north of India to Xinjiang (northwest China), is less clear. After Baktra, Ptolemy describes the next landmark as "the ascent of the range of the Kōmēdai," after which one reached "the gorge that follows upon the plains," and one ascended this gorge for fifty *schoinoi* (1,500 stades) to the Stone Tower. The description of these stages is rather obscure, but it suggests, first, an ascent to a plateau, followed by a further uphill journey along a gorge. From other information provided by Marinos, Ptolemy infers that the trends of these three stages were respectively northeast, southeast, and again northeast. In 6.13, Ptolemy gives the geographical data for the district inhabited by the nomads known as Sakai that contained the Stone Tower, and from these coordinates it would follow that the interval from Baktra to the ascent of the range was about 3,500 stades only slightly north of due east, from the ascent to the gorge was about 2,700 stades southeast, and from the beginning of the gorge to the Stone Tower was about 2,700 stades again northeast. (This last would be equivalent to ninety *schoinoi* rather than the fifty *schoinoi* mentioned in 1.12.)

There have been various attempts to identify these places and the specific route described by Marinos and Ptolemy. The name Tash Kurgan, which means "stone tower," belongs to more than one town in the broad district through which the route passed, though it need not follow that any of these sites was the Stone Tower of antiquity. Aurel Stein's attempt to locate the route deserves consideration, because it was based on personal experience of the local topography as well as documentary sources.[2] He proposes that it went northeast across the Oxus near Termez, Uzbekistan, up the broad valley of the river Surkhandarya, then south of east past modern Dushanbe, Tajikistan, and then northeast up the defile of the river Kyzylsu to the Stone Tower, which would have been approximately at the site of Daraut-Kurgan, Kyrgyzstan. He further suggests that the "base [*hormētērion*] of those who trade into Sēra at the Imaos range" would have been close to Irkeshtam, Kyrgyzstan, which is still a border post on the route to Kashi (Kashgar) in Xinjiang.

Ptolemy tells us that Marinos added up the stade distances from Hierapolis to the Stone Tower as 876 *schoinoi*, which Ptolemy reduces to a round 800. He

[2]Stein 1928, 2:848–850 and 893–894.

justifies this change in general terms by asserting that Marinos had not taken into consideration the deviations of the route from the parallel through Rhodes from Hekatompylos on, which are in fact the reason for his summarizing the stages of the journey. He does not, however, give us the numbers of stades belonging to each stage or attempt to show us just how much each stage ought to be reduced in accordance with its divergence from due east, and one suspects that the number he adopts is arbitrary. Using the standard equations of one *schoinos* with 30 stades and one degree along the parallel through Rhodes with 400 stades, Ptolemy's corrected value for the longitudinal interval from Hierapolis to the Stone Tower amounts to exactly 60°. The actual longitudes he assigns to the two places on his map differ by 65°.

For the remainder of the route, from the Stone Tower to the city of the Sēres, Marinos was compelled to estimate a stade distance from the report that the journey took seven months. Marinos derived this information from "one Maes, also known as Titianus, a Macedonian and a merchant by family profession" (1°.11), who had in turn learned of the duration of the journey from people that Ptolemy characterizes as Maes' agents. We can safely assume that the route from the Stone Tower descended into Xinjiang, and skirted the great Taklamakan desert (along its south side?) to central China. How much further it progressed depends on whether the "metropolis of the Sēres" was Lo-Yang (the Chinese capital at this period) or a less remote provincial city, a question on which it is perhaps pointless to speculate. It is also not clear just how Marinos obtained 36,200 for the number of stades along the parallel through Rhodes corresponding to this part of the journey; the assumed rate of travel would have had to have been in the neighborhood of 170 stades (approximately 30 kilometers) per day.

# *Appendix D*
# *The Breadth of the Mediterranean According to Marinos and Ptolemy*

In the last part of 1.12 Ptolemy reports, on Marinos' authority, the number of degrees of longitude between a succession of localities from the western limit of the *oikoumenē* to the crossing of the Euphrates near Hierapolis. In Table 1 we list these intervals, both in degrees (as Ptolemy presents them) and as fractions of an hour-interval (15°), and the running totals in degrees, which are of course the longitudes of the localities. For example, in the third row of the table, Marinos has placed the mouths of the Baetis River 2½° (i.e., ⅙ hour) east of the Sacred Cape, which gives the mouths of the Baetis a total longitude of 5° east of the Islands of the Blest. It can be seen that most of Marinos' intervals can be ex-

TABLE 1. Intervals along the Parallel through Rhodes According to Marinos and Ptolemy

| | Marinos | | | Ptolemy | |
|---|---|---|---|---|---|
| | Degrees | Hours | Longitude | Degrees | Longitude |
| Islands of the Blest | — | — | 0 | — | 0 |
| Sacred Cape | 2½ | ⅙ | 2½ | 2½ | 2½ |
| Mouths of the Baetis | 2½ | ⅙ | 5 | 2⅚ | 5⅓ (east mouth)[a] |
| Straits of Hēraklēs | 2½ | ⅙ | 7½ | 2⅙ | 7½ |
| Caralis | 25 | 1⅔ | 32½ | 25 | 32½ |
| Lilybaeum | 4½ | 3/10 | 37 | 4½ | 37 |
| Pachynus | 3 | ⅕ | 40 | 3 | 40 |
| Tainaros | 10 | ⅔ | 50 | 10 | 50 |
| Rhodes | 8¼ | 11/20 | 58¼ | 8⅓–8⅔ | 58⅓–58⅔ |
| Issos | 11¼ | ¾ | 69½ | 10⅔–11 | 69⅓ |
| Euphrates (Zeugma) | 2½ | ⅙ | 72 | 2⅔ | 72 |

[a]No western mouth is mentioned at 2.4.5 in the manuscripts of the *Geography*. The omission is older than Marcianus (Müller 1855–1861 1:545–546).

TABLE 2. Intervals along the Parallel through Rhodes According to Marinos and Strabo

| | Marinos | | Strabo | Actual |
|---|---|---|---|---|
| | Degrees | Stades | Stades | Stades |
| Sacred Cape | — | — | — | — |
| Straits of Hēraklēs | 5 | 2,000 | 3,000 | 1,800 |
| Pachynus | 32½ | 13,000 | 13,000+ | 10,000 |
| Rhodes | 18¼ | 7,300 | 7,500+ | 6,400 |
| Issos | 11¼ | 4,500 | 5,000 | 4,000 |
| Total | 67 | 26,800 | 28,500+ | 22,200 |

pressed as round fractions of hour-intervals, which may therefore have been the units that Marinos preferred for specifying longitudes. The last two columns give the number of degrees of longitude for each interval according to Ptolemy's own coordinates in Books 2–5 and Ptolemy's longitude for the eastern end of each interval. The comparison shows that Ptolemy adhered quite closely in his own map to Marinos' quantitative data for the Mediterranean.

Although it is likely that Marinos expressed these longitudes as time intervals between the meridians through the places in question, their empirical basis would almost certainly have been estimated stade distances between the places rather than the difference in time between observed times of lunar eclipses. Reports of simultaneous eclipse observations were, as Ptolemy tells us, hard to come by, nor would they have had the precision necessary to determine small longitudinal intervals.[1] On the other hand, stade distances reckoned along the parallel through Rhodes were part of the stock data of Greek geography, as can be seen from Strabo's extended discussion of such estimates (2.4.2–2.4.3).

Strabo cites stade distances spanning the Mediterranean, without specifying his sources but asserting that the figures were generally accepted in his time. In Table 2 we compare these with the numbers of stades that result from multiplying Marinos' longitudinal intervals by the standard factor of 400 stades per degree along the parallel through Rhodes, and with stade distances derived from the actual locations of the same places (assuming one stade = 185 meters).

Marinos' stade figures evidently reflect the same state of geographical knowledge as Strabo's, and their total is only about 20 percent in excess of the true distance. Expressed in degrees, however, the interval from the Sacred Cape to Issos is about 50 percent too large (67° instead of the correct 45°) because of the error in Marinos' assumption that 1° of the equator equals 400 stades.

[1]See pp. 29–30.

# *Appendix E*
# *The Sail to Kattigara*

Marinos' second estimate of the longitudinal extent of the *oikoumenē* made use of a series of reported distances along a trade route from Cape Kōry (at the straits separating Sri Lanka from India) to Kattigara. The journey appears to have followed the coast of the Bay of Bengal, taking a direct route east across the north end of the bay (see the Geographical Index s.v. Paloura and Sada), taking an overland shortcut across the Malay Peninsula, and then continuing along the coast of the Gulf of Siam as far as Zabai, a city near the southern tip of Vietnam (1.13). The final stage of the journey, from Zabai to Kattigara, was described by Marinos' source Alexandros as being several days' sail south and a bit leftward, i.e., eastward (1.14). If this was approximately correct, Kattigara would have to have been somewhere among the islands of Indonesia. Because Kattigara was allegedly a port for trade with the Sinai, i.e., the Chinese, it is usually supposed that Marinos and Ptolemy were mistaken about the direction, and that Kattigara was somewhere on the Vietnamese or Chinese mainland. It must be admitted, however, that Ptolemy speaks of this stage as a "sail across" (*diaplous*), a term that he reserves for a crossing of open water.

Ptolemy supplies us with the figures in stades that Marinos had for the stages of the voyage from Cape Kōry to the Golden Peninsula. Ptolemy presumes that these figures were obtained by multiplying the actual time of travel by an ideal number of stades sailed in a day; whereas Marinos apparently took the sum of the numbers as giving the longitudinal interval directly. Hence Ptolemy makes schematic reductions (1) to get the true distance traveled on the supposition that interruptions and delays would have led to a less-than-ideal progress in a day, (2) to get the straight-line distance on the supposition that distances across bays give the length of coast rather than the straight line route, and (3) to get the longitudinal component of intervals that were not due east. Degrees of longitude are obtained by dividing the stade figures by 500, i.e., for these latitudes close to the equator one neglects the curvature of the globe.

In this way, Ptolemy obtains an east-west interval of $34\frac{4}{5}$° from Cape Kōry to the Golden Peninsula. Without Ptolemy's reductions, Marinos would have found approximately 61⅙°. For the remaining interval to Kattigara, Marinos' procedure is not clear. According to Ptolemy, Marinos did not write down a figure in stades for the stage from the Golden Peninsula to Zabai, but stated that this was twenty days' sail along a southward-facing coast. Supposing that Marinos used the estimate of 1,000 stades for a day's sail, as he did when measuring the east coast of Africa, he would have had another 40° to add to the longitude. The final stage, from Zabai to Kattigara, was supposedly almost due south, so that even though Marinos attempted to interpret Alexandros' report as saying that this was a long sail, the contribution to the longitude cannot have been many degrees.

We may guess, then, that Marinos reckoned the longitudinal interval from Cape Kōry to Kattigara as a little more than 100°. Adding this to the interval of slightly over 125° from the Islands of the Blest to Cape Kōry that Ptolemy derives from Marinos' data, we obtain a total longitudinal extent of something over 225° for the whole *oikoumenē*, as we should expect.

In the process of revising Marinos' calculations, Ptolemy has incidentally determined the longitudes of Cape Kōry, Kouroula, Paloura, Sada, Tamala, and the terminus of the crossing of the Golden Peninsula at the inlet of the Bay of Perimoula. On his own map he respected these longitudes, making allowance for small deviations (only once above a degree). From the Bay of Perimoula to Zabai, however, he allows only 9⅙°, which is significantly less than the 10⅓° that he allots to this interval in his audaciously contrived calculation of the longitude of Kattigara—although Kattigara itself is set at 177°, in exact agreement with his conclusion in 1.14.[1]

[1]In all these longitudes Ptolemy appears to have forgotten the unmeasured small interval between the meridian through the sources of the Indus and Cape Kory.

# *Appendix F*
# *The Miscellaneous Criticisms of Marinos' Data*

In 1.15 Ptolemy tersely lists a series of inconsistencies that he has found in Marinos' geographical data. The following notes attempt to make the nature of the conflict clear where it is not obvious from Ptolemy's statement. We also make a comparison with the way that Ptolemy himself dealt with the localities in question, and with the actual geographical configuration of the localities. One gets the general impression that Ptolemy was eclectic in choosing which of Marinos' data—if any—to retain on his map, and that he by no means always made the correct choices.

The first group of Marinos' inconsistencies discussed in 1.15 involves localities that Marinos stated were "opposite" each other, i.e. (as Ptolemy has explained in 1.4), they lie on the same meridian.

> (1) He says... that Tarraco is opposite Caesarea Iol, although he draws the meridian through [Caesarea Iol] also through the Pyrenees, which are more than a little to the east of Tarraco.

According to Ptolemy's geographical catalogue, Tarraco has longitude 16⅓° (2.6.17), while Caesarea Iol is further east at longitude 17° (4.2.5), and the Pyrenees extend from 15⅙° (or 15° in manuscript **U**) to 20⅓°. There is, of course, no single meridian through the Pyrenees; Marinos may have meant the midpoint of the range, or its easternmost point. Ptolemy has not followed either of Marinos' indications. In actuality, the meridian of Caesarea Iol lies east of Tarraco, and passes close to the east end of the Pyrenees.

> (2) [He says that] Pachynus is opposite Leptis Magna, and Himera opposite Thēna, yet the distance from Pachynus to Himera amounts to about 400 stades, while that from Leptis Magna to Thēna amounts to over 1,500 [stades] according to what Timosthenes records.

Presumably it is the mouths of the river Himera on the south coast of Sicily, not the city and river of the same name on the north coast, that is intended, since the site is said to be "opposite" Thēna (Theaenae) on the north coast of Africa. Marinos' data would obviously be inconsistent if Himera is assumed to be roughly west of Pachynus on the Sicilian coast, and Thēna roughly west of Leptis Magna on the African coast.

Ptolemy's coordinates for the four places (3.4.7–3.4.8 and 4.3.11–4.3.13) lead to distances in close agreement with Marinos' figures. Himera and Pachynus are made to lie between the meridians through Thēna and Leptis Magna. This suggests that Ptolemy trusted the estimates of coastal distances more than the reported directions of sail.

Thēna really lay more than 2,000 stades in a straight line from Leptis, although the longitudinal component of this interval is only about 1,700 stades because Thēna was a fair bit north of Leptis, and not on the same parallel, as Ptolemy believed. Similarly, the longitudinal interval between Himera and Pachynus was almost 600 stades. Moreover, each of the localities in Sicily is actually east of the corresponding African city that it is said to be "opposite," so that both Marinos and Ptolemy have placed the entire island too far west relative to the African coast.

> (3) He says that Tergestē lies opposite Ravenna, whereas Tergestē is 480 stades from the inlet of the Adriatic at the river Tileventus in the direction of the summer sunrise, and Ravenna is 1,000 stades in the direction of the winter sunrise.

If Tergestē was really "opposite" Ravenna, and the mouth of the Tilaventus was really so situated that the directions to Tergestē and Ravenna could be symmetrically oriented with respect to due east, then the distances ought also to be equal. In 3.1.23–3.1.27 Ptolemy has kept Tergestē (at longitude 34½°) more or less north of Ravenna (at latitude 34⅔°), and places the mouth of the Tilaventus west of them at a latitude barely south of Tergestē. Both distances are between 500 and 600 stades.

Marinos' data are in fact entirely incorrect. Ravenna was southwest of Tergestē, and about 700 stades south-southwest of the mouth of the Tilaventus; while Tergestē was less than 300 stades almost due east of the river mouth.

> (4) He says that Chelidoniai lies opposite Canopus, and Akamas opposite Paphos, and Paphos opposite Sebennytos, where again he sets the stades from Chelidoniai to Akamas as 1,000, and Timosthenes sets those from Canopus to Sebennytos as 290.

The configuration is similar to (2), with two pairs of places assumed to form a rough parallelogram: Chelidoniai on the south coast of Asia Minor; Akamas east of this on the west coast of Cyprus; Sebennytos due south of Akamas on the north coast of Egypt; and Canopus further west on the Egyptian coast and due south of Chelidoniai. Paphos, south of Akamas on the coast of Cyprus, is mentioned only in order to establish the relative positions of Akamas and Sebennytos. It is evident that by "Sebennytos" Marinos meant the Sebennytic mouth of the Nile, not the inland town of Sebennytos. "Paphos" is presumably New Paphos, the harbor city founded in the late fourth century B.C.

On Ptolemy's map, Akamas is about 1,100 stades east and just slightly south of Chelidoniai, while the Sebennytic mouth is about 320 stades due east of Canopus. Ptolemy has thus preserved Marinos' distances; but the localities in Cyprus are shifted quite far west of the meridian through the Sebennytic mouth.

As it happens, Marinos was approximately correct in stating that Chelidoniai is "opposite" Canopus, and Paphos opposite the Sebennytic mouth; and the distance from Chelidoniai to Akamas is indeed about 1,100 stades, though in a southeast direction. The error of assuming that Akamas was close to due east of Chelidoniai (which might be Ptolemy's rather than Marinos'), together with the fact that Timosthenes' figure for the distance from Canopus to the Sebennytic mouth is less than half the correct distance, account for the inconsistency.

The remaining instances of conflicting data are more miscellaneous:

> (5) He says that Pisae is 700 stades from Ravenna, in the direction of the *Libonotos* [south-southwest] wind, but in the division of the *klimata* and of the hour-intervals he puts Pisae in the third hour-interval and Ravenna in the second [MSS: fourth].

This passage has long been recognized to be problematic. In his map Ptolemy has retained Marinos' distance and direction. (In actuality Pisae was close to 1,000 stades from Ravenna, in a direction slightly west of southwest.) Hence it must be the assignment of the two cities to their hour-intervals that he has rejected. The most plausible interpretation of the hour-intervals is that they were longitudinal zones bounded by meridians 15° apart, and counted from the western end of the *oikoumenē*. But this leads to two difficulties: first, that according to Ptolemy's coordinates both places are near the *western* side of the third hour-interval (30°–45°); and second, that if the fourth hour-interval was east of the third hour-interval, then there would be no patent inconsistency with Marinos' assertion that the route from Ravenna to Pisae was to the south and west. Hence Marinos' hour-intervals worked in a different way, or Ptolemy was confused, or our text of the passage is damaged.

Let us consider the first possibility. If the hour-intervals were counted from east to west, then the assignment of Ravenna to a higher-numbered interval than Pisae would be obviously inconsistent with the alleged direction. But why would Marinos have chosen a prime meridian 60° east of north Italy, i.e., at about 79° east of the western limit of the *oikoumenē* and passing through, say, Mesopotamia? Still less credible is the notion that, in this context alone, "hour-interval" means a *latitudinal* belt between parallels along which the longest day of the year is a whole number of hours.[1] This would in any case be little help, since the only way to get an inconsistency with the direction from Ravenna to Pisae would be to count these "hour-intervals" *southward*, and not even from the northern limit of the *oikoumenē* (63° N) but from the insignificant parallel 61° N. Thus we conclude that the hour-intervals were indeed counted from west to east.

The possibility that Ptolemy misunderstood Marinos need not detain us. It is scarcely to be believed that Ptolemy saw an inconsistency where there was none in such a simple geographical situation, or (as Honigmann has supposed) that he failed to realize that the Pisae in Marinos' section on the hour-intervals was a different town in Greece.[2] The likeliest resolution, therefore, is to suppose that the text of 1.15 has not been transmitted accurately. And in fact, all that we need to assume is that Ptolemy wrote that Marinos set Ravenna in the *second* hour-interval, effectively giving it a longitude just under 30° from the western limit. An early copyist could have substituted "fourth" through either absent-mindedness or misreading of a numeral.

> (6) Though he has said that Londinium in Britain is fifty-nine [Roman] miles north of Noviomagus, he then represents [Noviomagus] as north of [Londinium] in his division of the *klimata*."

Ptolemy's latitude for Noviomagus is not secure because of variant readings in the manuscripts, but the narrative of 2.3 makes it clear that Noviomagus is supposed to lie to the south of Londinium. The significant point seems to be that the parallel along which the longest day is seventeen equinoctial hours, which marked the boundary between two of Marinos' *klimata*, passes close to both places. Marinos apparently put Londinium in the *klima* south of the parallel, and Noviomagus in the *klima* north of it. His distance of fifty-nine Roman miles is very accurate for the straight-line distance (which is in a northeast direction).

[1]See also the refutation of one version of this hypothesis in Honigmann (1930) 1782–1783.

[2]Honigmann (1930) 1783. Since Ravenna is in the third hour interval on Ptolemy's map, and this Olympia Pisae (III 16.18) in the fourth interval, Honigmann is forced to suppose, on top of Ptolemy's error, that the numerals were transposed in the manuscript tradition.

(7) Having put Athōs on the parallel through the Hellespont, he puts Amphipolis and its surroundings, which lie north of Athōs, and the mouths of the Strymōn, in the fourth *klima*, which is below the Hellespont."

Ptolemy (3.13) places Amphipolis and the mouths of the Strymōn a little north of Athōs, and all of them north of the parallel through the Hellespont (40¹¹⁄₁₂°, 1.23.13). The statement that the *klima* south of this parallel is the fourth is consistent either with the hypothesis that Marinos' *klimata* were bounded by the parallels at half-hour increments of longest day, starting with the parallel through Meroē (longest day equals thirteen hours), or with the hypothesis that they were bounded by the parallels at quarter-hour increments, starting with the parallel through Alexandria (longest day equals fourteen hours).

Marinos was in fact correct in placing Athōs on the same parallel as the Hellespont; but the other localities are, as Ptolemy says, further north.

(8) Although almost the whole of Thrace lies below the parallel through Byzantion, he has set all Thrace's inland cities in the *klima* above this parallel.

In fact, most of Thrace *is* north of the latitude of Byzantion, but Ptolemy was misled by the traditional latitude of 43¹⁄₁₂° assigned to Byzantion, which is about 2° too far north. Consequently Ptolemy disregards the information in Marinos' section on the *klimata*, placing about three-fifths of the inland cities of Thrace south of the parallel.

The parallel through Byzantion, which is said to be the southern boundary of a *klima*, is that along which the longest day is 15¼ hours—supporting the hypothesis of quarter-hour increments between Marinos' *klimata*.

(9) He says: "We shall situate Trapezous on the parallel through Byzantion;" and after showing that Satala in Armenia is sixty miles south of Trapezous, nevertheless in the description of the parallels he puts the parallel through Byzantion through Satala, and not through Trapezous.

Ptolemy rightly accepts Marinos' first assertion (from his section on the *klimata*?) that Trapezous lies on the parallel through Byzantion (5.6.5), and places Satala about a degree south of this parallel (5.7.3).

(10) He even says that the river Nile, from where it is first seen up to Meroē, will be drawn correctly [going] from south to north. Likewise, he says that the sail from Arōmata to the lakes from which the Nile flows is effected by the *Aparktias* [north] wind. But Arōmata is quite far east of the Nile; for

> Ptolemais Thērōn is east of Meroē and the Nile by a march of ten or twelve days, and <the Bay of Adoulis is... stades> from Ptolemais, and the straits between the peninsula of Okēlis and Dērē are 3,500 stades from Ptolemais, and the cape of Great Arōmata is 5,000 stades to the east of these.

Ptolemy understood Marinos as saying that the lakes that were the sources of the Nile lay quite close to the eastern coast, so that one could speak of sailing due south from Arōmata *to* these lakes. Marinos presumably derived this notion from Diogenes' report of his storm-driven voyage (1.7). Ptolemy argues that Arōmata must lie even further away from the Nile than Ptolemais Thērōn, so that if it is accepted as a hypothesis that the Nile's course is due south, either the coast must tend westward south of Arōmata or the lakes must be far east of the coast.

In spite of some uncertainty about just how one is meant to draw the rivers near Meroē, the longitudes in Ptolemy's catalogue (4.7) show that Ptolemy has kept the course of the river south of Meroē close to 61° longitude, as Marinos apparently had done. The directions for the places along the coast (4.7 and 4.10) accord with what Ptolemy says in 1.15, as do the distances from Meroē overland to Ptolemais and from Dērē to Arōmata. From this last interval, which is not even nearly due east, it is evident that Ptolemy understands the stated distances as applying to the total journey, not just to the east-west component. The 3,500 stades to Dērē clearly must be counted from the beginning of the bay of Adoulis, not from Ptolemais, and we therefore infer that another eastward stage from Ptolemais to the bay of Adoulis has dropped out of our text of I 15 through a skip of the scribe's eye. Beyond Arōmata, Ptolemy has the coast tend southwest as far as Rhapta, though it still does not approach very near the lakes. Broadly speaking, Ptolemy's outline of the coast and his course for the river are a fairly good approximation of the truth.

# *Appendix G*
# *Textual Notes*

We here list passages where we have translated a version of the text significantly different from that in the edition of Nobbe, either because a superior reading is transmitted in the manuscripts, or because we have adopted a conjectural correction of the transmitted text. The passages are specified by the chapter and section number according to Nobbe's division; we also provide his volume, page, and line numbers in parentheses. In 2.7–2.10, we have followed the superior text printed by Cuntz, and have not reported his divergences from Nobbe.

For the most part we have found it sufficient to report the readings of the two manuscripts **X** and **U**, as representatives of the two principal recensions of the text. These are not complete collations, and we have ignored variant readings that do not affect the translation or that are obviously false, as well as the many places where we have altered Nobbe's punctuation.

## *Book 1*

1.1.1 (1:3.3) διαγραφῆς **U**, Nobbe: διὰ γραφῆς **X**, our text.

1.1.4 (1:4.12) ἐπιπλεῖον **U**, Nobbe: ἐπὶ πλεῖον **X**: deleted from our text (it is probably an intrusion from ἐπὶ πλεῖστον in 1.1.5).

1.1.5 (1:4.13) ἐπιπλεῖστον **XU**, Nobbe: ἐπὶ πλεῖστον our text.

1.1.6 (1:4.23) γραμμάτων **U**, Nobbe: γραμμῶν **X**, our text.

1.1.9 (1:5.7) ἐπιδεικνύντα **U**, Nobbe: ἐπιδεικνῦναι **X**, our text (it is not the data discussed in the preceding sentences that exhibit the true nature of the heavens and the earth, but the subject to which they belong).

1.3.1 (1:8.26) τῶν μεσημβρινῶν **U**, Nobbe: τοῦ μεσημβρινοῦ **X**, our text.

1.3.4 (1:10.5) καὶ ἔτι τὴν ἀπολαμβανομένην τοῦ ἰσημερινοῦ περιφέρειαν ὑπὸ τῶν δύο μεσημβρινῶν, ἐὰν ἕτεροι ὦσι τοῦ ἰσημερινοῦ παράλληλοι Nobbe: καὶ ἔτι τὴν ἀπολαμβανομένην ὑπὸ τῶν δύο μεσημβρινῶν ἐὰν ἕτεροι ὦσι τοῦ ἰσημερινοῦ παράλληλοι **U**: καὶ ἔτι τὴν ἀπολαμβανομένην περιφέρειαν τοῦ ἰσημερινοῦ ὑπὸ τῶν δύο μεσημβρινῶν ἐὰν ἕτεροι ὦσι τοῦ ἰσημερινοῦ **X**: καὶ ἔτι τὴν ἀπολαμβανομένην ὑπὸ τῶν δύο μεσημβρινῶν, ἐὰν ἕτεροι ὦσι, τοῦ ἰσημερινοῦ our text (the additional words in **U** and in Nobbe result from failing to see that ἐὰν ἕτεροι ὦσι is parenthetic).

1.7.2 (1:14.20) τουτέστι **XU**, Nobbe: deleted from our text.

1.7.4 (1:14.29) ὅλος **XU**, Nobbe: deleted from our text.

1.9.3 (1:19.18) μόνῳ Nobbe: μόνον **XU**, our text.

1.9.7 (1:20.25) οὐ Nobbe: ἢ **XU**, our text.

1.12.1 (1:25.14) μὴ Nobbe: deleted from our text.

1.12.6 (1:26.23) ἀπὸ τῆς ὁμωνύμου πόλεως, ἥ ἐστιν ὀλίγῳ βορειοτέρα Nobbe: ἀπὸ τῆς ὁμωνύμου πόλεως ὀλίγῳ βορειοτέραν **U** (uncorrected): ἃ τῆς ὁμωνύμου πόλεως ὀλίγῳ βορειότερα **U** (corrected): ἃ τῆς ὁμωνύμου πόλεώς ἐστιν ὀλίγῳ βορειότερα **X**, our text.

1.15.5 (1:33.29) πρὸς λιβόνοτον **U**, Nobbe, our text: om. **X**.

1.15.5 (1:34.2) τετάρτῳ **XU**, Nobbe: δευτέρῳ our text. For an explanation of our emendation, see Appendix F, pp. 139–140.

1.15.11 (1:34.25) θηβῶν **X**, Nobbe: θηρῶν **U**, our text.

1.15.11 (1:34.26–27) Πτολεμαΐδος δὲ καὶ τοῦ 'Αδουλικοῦ ('Αδουλιτικοῦ **X**) κὸλπου **XU**, Nobbe: Πτολεμαΐδος δὲ ⟨ὁ 'Αδουλιτικῷ κόλπῳ σταδίοις *x*⟩, καὶ τοῦ 'Αδουλιτικοῦ κὸλπου our text (*x* stands for a lost numeral). See Appendix F, pp. 141–142.

1.17.7 (1:37.8) αἰγιαλοῦ **U**, Nobbe: αἰγιαλῶν **X**, our text.

1.17.12 (1:38.1) τὸ καλούμενον Νίκι **U**, Nobbe: τὸ καλούμενον Τονίκι **X**, our text. In 4.7.11, the name is rendered Τονίκη (Nobbe), Τονίκι (**X**), Τωνίκι (**U**).

1.18.1 (1:38.12) ἀλλῷ ἵνα μὴ **U**, Nobbe: μὴ καὶ **X**, our text.

1.18.1 (1:38.13) ἔσται Nobbe: καὶ γὰρ**X**: ἔσται γὰρ**X** corr., **U**, our text.

1.20.6 (1:42.7) περιφορᾶς Nobbe: παραφορᾶς **XU**, our text.

1.23.1 (1:45.19–22) γεγράπται... πρὸς τὰς ἄρκτους **XU**, Nobbe: transposed to the end of the chapter in our text.

1.23.7 (1:46.9) καὶ γραφόμενον διὰ Συήνης **U**, Nobbe, our text: om. **X.**

1.23.9 (1:46.11) τὸν καὶ δι' 'Αλεξανδρείας Nobbe: om. **XU**, our text.

1.23.13 (1:46.18) τὸν καὶ δι' Ἑλλησπόντου γραφόμενον Nobbe: om. **XU**, our text.

1.23.15 (1:46.21) καὶ διὰ μέσου Πόντου Nobbe: om. **XU**, our text.

1.23.16 (1:46.23) τὸν καὶ διὰ Βορυσθένους Nobbe: om. **XU**, our text.

1.24.3 (1:49.4) ὁμοίων **XU**, Nobbe: μοιρῶν our text.

1.24.3 (1:49.9) ι α Nobbe: om. **XU**, our text.

1.24.6 (1:49.26) σλυ Nobbe: ΣΥ **XU**, our text.

1.24.18 (1:55.19) ζ θ ο Nobbe: Ξ Θ Ο **XU**, our text.

1.24.21 (1:56.9) τοῦ Θ δ **XU**, our text: τοῦ δ Nobbe.

1.24.27 (1:57.17) ζ η **X**, Nobbe: ΖΚ **U**, our text.

1.24.27 (1:57.17–18) ἡ θ ω ἐλάττων ἔσται τῆς πρὸς τὴν κ ζ συμμετρίας, καθάπερ καὶ τῆς [θν] θ τ Nobbe: ἡ ΘΩ ἐλάττων ἔσται πρὸς τὴν ΖΚ συμμέτρου, καθάπερ καὶ τῆς ΘΤ **X**: ἡ ΘΩ ἐλάττων ἔσται πρὸς τὴν ΚΖ συμμέτρου, καθάπερ καὶ τῆς ΘΥ ΥΤ **U**: ἡ ΘΩ ἐλάττων ἔσται τῆς πρὸς τὴν ΚΖ συμμέτρου, καθάπερ καὶ τῆς ΘΤ our text.

## *Book 2*

2.1.2 (1:61.17) κατὰ συνεγγισμὸν τῶν πρὸς τὸ ἀξιοπιστότερον εἰλημμένων θέσεων ἢ σχηματισμῶν **U**, Nobbe: κατὰ συνεγγισμὸν τῶν πρὸς τοὺς ἀξιοπιστότερον εἰλημμένους θέσεων ἢ σχηματισμῶν **X**: κατὰ συνεγγισμὸν αὐτῶν πρὸς τοὺς ἀξιοπιστότερον εἰλημμένους θέσεις ἢ σχηματισμούς our text.

2.1.6 (1:62.22) διὰ **XU**, Nobbe: ἀπὸ our text.

2.1.7 (1:63.1) ὅλην **XU**, Nobbe: ⟨πρὸς⟩ ὅλην our text.

## *Book 7*

7.5.1 (2:176.20–177.2) In **X** this paragraph is displaced to immediately before 7.7, where it is preceded by the heading, ⟨τ⟩ὰ προγραφόμενα πρὸ τῆς καιφαλαιώδης [*sic*] ὑπογραφῆς τῆς οἰκουμένης καὶ τῶν ἐπαρχιῶν, "the things that are written before the concise caption of the *oikoumenē* and the provinces."

7.5.1 (2:176.23) τοῖς καθ' ἕκαστον, ὡς ἐν ἱστορίᾳ Nobbe: τῆς καθ' ἕκαστον, ὡς ἐν ἱστορίᾳ **U**: τῶν καθ' ἕκαστον, ὡς ἐνῆν ἱστορίᾳ **X**: τῶν καθ' ἕκαστον ὡς ἐνῆν μάλιστα our text.

7.5.16 (2:180.26) ὡσαύτως καὶ ἐν τῷ διὰ Μερόης, ἐν δὲ τῷ ἰσημερινῷ ρμ [*sic* Nobbe, om. **U**] ιβʹ **U**, Nobbe: om. **X**, our text.

7.5.16 (2:180.29) καὶ ἔτι **XU**, Nobbe: ὥστε our text.

7.6.1 (2:181.6) καταγράφοιτο **U**, Nobbe: διαγράφοιτο **X**, our text.

7.6.2 (2:181.13) ἡ Nobbe: om. **XU**, our text.

7.6.2 (2:181.18) τῆς γῆς Nobbe: ἐγγης **X**: ἔγγιστα **U**, our text.

7.6.3 (2:182.6) μέσου **U**, Nobbe: μέσων **X**, our text.

7.6.7 (2:184.2) ἔσται **XU**, Nobbe: ἔστω our text.

7.6.7 (2:184.8) ε ο Nobbe: ΕΣ **XU**, our text.

7.6.7 (2:184.9) ξγ, ἡ δὲ ε γ Nobbe: κγ L γ, ἡ δὲ ΕΤ **XU**, our text.

7.6.9 (2:184.21) κατὰ Nobbe: καὶ **XU**, our text.

7.6.9 (2:184.27) τὸ Nobbe: τὰ **XU**, our text.

7.6.9 (2:184.29) πέντε Nobbe: πέρατα **XU**: περὰν our text.

7.6.10 (2:185.7) (τὰ) τῶν (εἰρημένων) **XU**, our text: om. Nobbe.

7.6.10 (2:185.7–8) α τ μ καὶ γ λ δ Nobbe: ,αΤ,β καὶ τὸ ,γ⟦illegible letter⟧,δ **U**: ,αΤ,β καὶ ,γΥ,δ **X**, our text.

7.6.11 (2:185.11) ἐπὶ **XU**, Nobbe: ἔτι our text.

7.6.11 (2:185.14–15) ς ζ η καὶ θ καὶ ε Nobbe: ,ε,ζ,ν καὶ τὸ ,θκ,ς καὶ τὸ ,ε **X**: ,εχ,ζη καὶ τὸ ,θ καὶ ,ς **U**: ,ε,ζ,η καὶ τὸ ,θ,κ,λ our text.

7.6.13 (2:187.3) ἐκεῖνα **XU**, Nobbe: deleted from our text.

7.6.13 (2:187.4) εἰρημένων Nobbe: εἰλημμένων **XU**, our text.

7.6.13 (2:187.6) τῇ... τομῇ **U**, Nobbe: ταῖς... τομαῖς **X**, our text.

7.6.13 (2:187.7) κατάματος Nobbe: κατάγματος **XU**, our text.

7.6.15 (2:187.23) ἐπὶ Nobbe: παρὰ **X**: περὶ **U**, our text.

7.7.3 (2:189.26) ἐκκειμένην μένει Nobbe: ἐκκειμένην **XU**: ἐκείνην our text.

7.7.3 (2:189.27) κύκλοι Nobbe: τῶν ἐντεταγμένων κύκλων **XU**, our text.

7.7.3 (2:190.1) οἱ μὲν τῆς μεσημβρινῆς Nobbe: οἱ μὲν τῆς μεσημβρινῆς πρὸς τὴν **XU**: οἱ μὲν μεσημβρινοὶ πρὸς τὴν our text.

7.7.3 (2:190.5) ἀνακεκλιμένος **U**, Nobbe: ἀνακεκριμένος **X**, our text.

7.7.4 (2:190.9) μὴ περιρρέοντος Nobbe: περιρ⟨ρ⟩έοντος αὐτὸ (αὐτῷ **X**) **XU**, our text.

## *Book 8*

8.1.5 (2:194.7) εἱς ἓν Nobbe: om. **XU**, our text.

8.2.1 (2:195.9) πόσος Nobbe: ποστός **XU**, our text.

8.5.3 (2:200.10–12) Longitude of Mediolanium: β γ′ Nobbe: β ⌐ο **X**: β ∟ γ′ **U**, our text.

8.5.5 (2:200.17–19) Augustodunum, latitude: ιε ∟ δ′ **U**, Nobbe: ιε ∟ γ′ **X**, our text. Longitude: β γ′ ιβ′ **U**, Nobbe: β ∟ **X**, our text.

8.5.6 (2:200.24–26) Gesoriacum, longitude: β ∟ δῳ′ Nobbe: β ∟ **XU**, our text.

8.5.6 (2:200.27–29) Durocottorum, longitude: β γ′ ιβ′ **U**, Nobbe: β ∟ **X**, our text.

8.5.7 (2:201.10–12) Nemausus, longitude: β ∟ ιε′ **U**, Nobbe: β ∟ **X**, our text.

# *Appendix H*
# *Geographical Index*

*This index lists all localities mentioned in the chapters of the Geography translated above, except those that appear only in 2.4–7 and 8.5. An asterisk before a place name indicates that it has a separate index entry.*

Adoulis, Bay of [Map 7, 1.15]: mod. Gulf of Zula, on the African coast of the Red Sea. The town of Adoulis, after which the bay was named, was apparently by the harbor of mod. Mits'iwa (Massawa). See Casson (1989) 102–106.

Adriatic Bay [Map 5, 1.15, 7.5]: mod. Adriatic Sea, which Ptolemy treats as a bay of the *Mediterranean.

Aegean Sea [Map 6, 7.5].

Agisymba [Map 7, 1.7–12, 7.5]: district in the interior of the continent of *Libyē, at the southernmost limit of the *oikoumenē* as known to Marinos and Ptolemy; the name appears in no other independent ancient source. According to 1.8, it was reached by Julius Maternus after a journey of four months southward from *Garamē, and the king of Garamē claimed the *Aithiopians who inhabited Agisymba as his subjects. In 4.8.5 Ptolemy describes Agisymba as a mountainous country of great extent, and in 1.8 we are told that this is where rhinoceros have their gathering place. The location of Agisymba cannot be established with any certainty, but it was probably not nearly so far south as Marinos and Ptolemy believed: perhaps one of the mountainous Saharan districts of Air (Niger) or Tibesti (Chad) near latitude 20° N. For discussion, see Desanges 1978, 197–200.

Aithiopia [Map 7, 1.8, 2.1, 7.5]: the parts of the continent of *Libyē south of Inner *Libyē and *Egypt.

Aithiopian Bay [Map 7, 7.5]: the bay on Ptolemy's map formed by the west coast of the continent of *Libyē where its trend changes from southward to westward; in 4.6.3–4.6.7 and 4.8.1–4.8.3 it is called the "Western" or "Great" Bay (not to be confused with the *Great Bay of Asia). It has no identifiable counterpart in the actual geography of Africa.

Aithiopians [1.7–10]: By this term Ptolemy means "dark-skinned people," as is clear from the argument of 1.9. Various peoples characterized as Aithiopians are listed in 4.6–8 as dwelling in Inner *Libyē (the part of the continent of *Libyē south of the Roman provinces of Mauretania, Africa, and Cyrenaica) and *Aithiopia. In 7.3.1–3 Ptolemy mentions "Fish-eating Aithiopians" as inhabiting the Bay of *Sinai.

Akamas [Map 6, 1.15]: mod. Cape Akamas (or Arnauti) in northwest Cyprus.

Albiōn [Map 4, 7.5]: the island of Great Britain.

Alexandria [Maps 5 and 6, 7.5, 8.2]: mod. Alexandria (El-Iskandariya), Egypt. Ptolemy uses the meridian through Alexandria as a reference for longitudes in the captions to the world and regional maps.

Amphipolis [Map 6, 1.15]: city in ancient Macedonia, mod. Amfipolis (Greece), near the mouth of the *Strymōn.

Antiocheia Margianē [Map 8, 1.12]: chief city of Margianē, a district (satrapy) of the Parthian Empire centered on the oasis of Mary (or Merv), Turkmenistan.

Arabia [1.7, 17, 2.1]: three distinct districts have this name in the *Geography*: Arabia Petraea (described in 5.17), Arabia Deserta (5.19), and *Arabia Felix (6.7). The Arabia that adjoins Egypt in II 1 is Petraea; the two other references are to Arabia Felix.

Arabia Felix [Map 7, 1.17]: Ptolemy applies this term to the whole of the Arabian peninsula.

Arabia, Bay of [Map 7, 2.1, 7.5]: mod. Red Sea.

Arabia, land strait [7.5]: see *Hērōopolis.

Arbēla [Map 8, 1.4]: city in Assyria (modern Arbīl, Iraq), near which Alexander the Great defeated Darius III in the battle of Gaugamela (or Arbēla) on October 1, 331 B.C. (The date of the battle, which has been disputed, is now determined exactly by the report in a contemporary Babylonian astronomical Diary.) See pp. 29–30.

Argarou, Bay of [Map 8, 1.13]: bay between Cape *Kōry and *Paloura, corresponding to the mainland coast of the modern Palk Strait between India and Sri Lanka.

Areia [Map 8, 1.12]: district (satrapy) about Herāt, Afghanistan.

Armenia [Map 6, 1.15]: not coextensive with the modern country. Ptolemy distinguishes between Lesser Armenia (5.7) west of the river Euphrates, and Greater Armenia (5.13) east of the river. *Satala is in the former.

Arōmata (or Great Arōmata) [Maps 6 and 7, 1.9, 14–15, 17]: cape at the Horn of Africa (Raas Caseyr, Somalia), where there was a trading post (*emporion*). The name means "spices." See Casson 1989, 129.

Asia [Map 6, 2.1, 7.5, 8.1–2]: the continent, called Great Asia in 7.5 to distinguish it from the Roman province of Asia (5.2).

Assyria [Map 8, 1.12]: district around the upper Tigris River in modern Iraq.

Athōs [Map 6, 1.15]: mod. Mt. Athos, Greece.

Azania [Map 7, 1.7, 9, 17]: East Africa between *Arōmata and *Rhapta. In 4.7 Ptolemy reserves the name for the district inland from the coast (which he there calls *Barbaria). See Casson 1989, 136.

Baetis River [Map 5, 1.12, 14]: mod. Guadalquivir River, Spain.

Baktra [Map 8, 1.12]: principal city in *Baktria, the medieval Balkh, northwest Afghanistan.

Baktria [Map 8, 1.17]: district (satrapy) of central Asia, south of the river Oxus (Amudar'ya) and north of the Hindu Kush (Afghanistan).

Barbaria [Map 7, 1.17]: according to Ptolemy, the traders' name for the east coast of Africa south of *Arōmata (cf. *Azania).

Beach, Great and Little [Map 7, 1.17]: see *Bluff.

Blest, Islands of the (or Fortunate Islands) [Map 7, 1.11–12, 14, 7.5]: by Ptolemy's time, the name originally belonging to the mythical home of the "blessed" dead had been applied to the Canary Islands. As the westernmost locality on Ptolemy's and Marinos' maps, they supplied Ptolemy's prime meridian.

Bluff [Map 7, 1.17]: Ptolemy's account of the sequence of bluffs and beaches along the coast of *Barbaria is essentially correct (see Casson 1989, 136–137). The Somali coast between Raas Macbār and Raas Illig, i.e. between about 9°30' N and 7°30' N, consists of a sequence of bluffs. Chapter 15 of the *Periplus* likewise mentions, beyond Opōnē, the Small and Great Bluffs of *Azania as extending for six "runs" (i.e., a sail of three days and nights). South of this point, approximately as far as Warshiik (2°18' N), the coast is beach, which like Ptolemy, the *Periplus* divides into a "Great" and a "Small," with a total sail of six runs.

Britain [1.15, 7.5]: the British Isles, comprising the two islands of *Albiōn (Great Britain) and *Hibernia (Ireland).

Byzantion, parallel through [1.11–12, 15]: Byzantion was a town in Thrace, on the European shore of the Bosporus (later Constantinople, mod. Istanbul). The parallel through Byzantion, at 43½° N (3.11.5), is that along which the longest day is 15¼ equinoctial hours.

Caesarea Iol [Map 5, 1.15]: on the coast of Mauretania Caesariensis, mod. Cherchel (Algeria).

Calpe [Map 5, 1.12]: Gibraltar. See Straits of *Hēraklēs.

Canopus (Kanōbos) [Map 6, 1.15]: town at the westernmost of the ancient mouths of the *Nile.

Caralis [Map 5, 1.12]: mod. Cagliari, Sardinia.

Carthage (Karchēdon) [Map 5, 1.4], north of mod. Tunis, Tunisia.

Caspian Gates [Map 8, 1.12]: a defile, east of mod. Tehran but not identifiable with certainty, that gave a passage through the Elburz Mts. of northern Iran, south of the Caspian Sea.

Caspian Sea: see *Hyrkanian Sea.

Chelidoniai [Map 6, 1.15]: mod. Beş, island near mod. Cape Gelidonya, southwest Turkey.

Corsica [Map 5, 7.5] (France).

Crete [Map 6, 7.5] (Greece).

Cyprus [Map 6, 7.5].

Dalmatia [Map 5, 1.16]: Roman province along the east coast of the *Adriatic.

Dērē [Map 7, 1.15]: town on the African side of the straits at the south end of the Red Sea; see *Okēlis.

Duēcalidonian Ocean [7.5]: see *Ocean.

Egypt [Map 7, I1.1]: on Ptolemy's map, Egypt is the habitable district around the *Nile and its delta, extending as far south as the First Cataract, and east to the Red Sea.

Ekbatana [Map 8, 1.12]: principal city of *Mēdia, modern Hamadān (Iran).

Essina [Map 7, 1.17]: trading post of uncertain location, probably on the coast of Somalia near Warsheik (see *Bluff, *Sarapiōn).

Euphrates, crossing of, at Hierapolis [1.11–12]: probably referring to a bridge across the river Euphrates linking Zeugma and Apamea in (ancient) Syria, near mod. Birecik, Turkey. The crossing was actually about 60 kilometers north of *Hierapolis.

Europe [Map 4, I1.1, 7.5, 7, 8.1–2]: on Ptolemy's map, the continent of Europe is divided from *Asia (from south to north) by the Aegean, the Black Sea, and the Sea of Azov, the river Don, and an arbitrary meridian drawn from the source of the Don to the northern boundary of the map. It also includes most of the major islands of the Mediterranean (except Cyprus) and the British Isles.

Fortunate Islands: see Islands of the *Blest.

Ganges, Bay of [Map 8, 1.13, 7.5]: mod. Bay of Bengal.

Garamaioi [1.12]: a people inhabiting central *Assyria (6.2).

Garamē [Map 7, 1.8–12]: city of Inner *Libyē, near mod. Jarmah, Libya.

Golden Peninsula (or Golden Chersonese) [Map 8, 1.13–14, 17, 7.5]: the Malay Peninsula. From Ptolemy's description (7.2.5) it appears that the name was applied only to the part of the peninsula south of what he calls Takōla (not identifiable) on its west coast and the inlet of the Bay of Perimoula (the Gulf of Thailand) on its east coast.

Gorge that follows upon the plains [Map 8, 1.12]: see *Kōmēdai.

Great Arōmata, cape [1.15]: see *Arōmata.

Great Bay [Map 8, 7.5]: a large northward-pointing inlet of the Sea of *India, bounded on the west by a cape, perhaps mod. Mui Bai Bung (or Cau Mau), just east of *Zabai, and on the east by the Southern Cape on the coast of the country of the *Sinai. The actual coast of southeast Asia has no such large bay at this point, and a sailing course setting out from this point northeast along the mainland would never turn back south in the way shown on Ptolemy's map. The problem of identifying the Great Bay is tied to that of *Kattigara, which Ptolemy situates further south on the coast of the country of the Sinai. If Kattigara was on the Asian mainland, somewhere in northern Vietnam or southern China, then the Great Bay could be equated with the Gulf of Tongking. If, on the other hand, Kattigara was in the East Indies, the Great Bay would turn out to be a geographical fiction, representing the South China Sea as if surrounded by unbroken coastline.

Hekatompylos [Map 8, 1.12]: capital of *Parthia, perhaps near Jajarm, Iran.

Hellespont, parallel through [1.11–12, 15–16]: 40 11/12° N, along which the longest day is 15 equinoctial hours.

Hēraklēs, Straits of [Map 5, 1.12, 2.1, 7.5]: mod. Straits of Gibraltar.

Hērōopolis, inlet at [Map 7, 2.1]: mod. Gulf of Suez.

Hibernia [Map 4, 1.11, 7.5]: Ireland.

Hierapolis [Maps 5 and 7, 1.11–12]: city in Syria, also called Bambykē, southwest of the meeting of the Sajur River and the Euphrates.

Himera [Map 5, 1.15]: mod. Licata, Sicily.

Hyrkania [Map 8, 1.12]: district (satrapy) along the southeast coast of the Caspian Sea (near the river Gorgān), and its principal city.

Hyrkanian (or Caspian) Sea [Map 8, 1.12, 7.5]: the Caspian Sea, but also conflated by Ptolemy with the Aral Sea, into which the Oxus (Amudar'ya) and Iaxartes (Syrdar'ya) rivers flow.

Imaon (Imaus) range [Map 8, 1.12, 16]: the range forms part of the northern boundary of *India, from the southeastern corner of the country of the *Sogdians (125°, 38½° N) along the country of the *Sakai to its southeast corner (123°, 35° N), and then turns almost straight northward to the northern limit of the *oikoumenē*. It thus corresponds in part to the Himalayas, in part to the Pamirs and Tian Shan.

India [Map 8, 1.7, 9, 13, 16–17, 8.1]: Ptolemy's "India This Side of the Ganges" (7.1) is the region bounded on the west by the *Indus and its tributaries, on the north by the *Imaon Range, and on the east by the Ganges. "India Beyond the Ganges" (7.2) extends to the east of this as far as the *Great Bay, where it adjoins the country of the *Sinai.

India, Sea of [Map 8, 7.5, 8.1]: the Indian Ocean.

Indus River [Map 8, 1.14, 17].

Issos [Map 6, 1.12]: the Gulf of İskenderun (Alexandretta).

Italia [Map 5, 1.16]: the Italian peninsula, not exactly coextensive with the modern country.

Judaea [Map 6, 2.1]: in the geographical catalogue (5.16) the province is called "Palaestina or Judaea Syria."

Kattigara [Map 8, 1.11, 13–14, 17, 23]: trading station for commerce between the west and the *Sinai. Kattigara is the last named place along the Indian

Ocean on Ptolemy's map, at the end of a long stretch of coast that bears little resemblance to the actual outline of southeast Asia. Its identity has, not surprisingly, been the subject of much controversy; the most common hypothesis is that it was near the modern Hanoi. See Appendix E for further discussion.

Kolchoi, Bay of [Map 8, 1.13]: the bay formed by the mainland coast of the Gulf of Mannar, between India and Sri Lanka.

Komaria, Cape [Map 8, 1.17]: mod. Cape Comorin, in actuality the southernmost point of India (though on Ptolemy's map *Paloura is further south).

Kōmēdai [Map 8, 1.12]: mountains within the country of the *Sakai, corresponding to part of the Pamirs. The "gorge" leading to the *Stone Tower was evidently one of the high valleys that characterize the region.

Kōry, Cape [Map 8, 1.13–14]: the mainland cape at the Pamban Channel between India and Sri Lanka.

Kouroula [Map 8, 1.13]: in the Bay of *Argarou (7.1.12), perhaps near Atirampattinam in Palk Strait.

Lakōnia [Map 6, 1.12]: district of the Peloponnese in Greece.

Leptis (or Lepcis) Magna [Map 7, 1.8, 10, 15]: city on the Mediterranean coast of the Roman province of Africa, near mod. Al Khums, Libya.

Libyē [Map 7, 2.1, 7.5, 7, 8.1–2]: the Greek geographical term for the continent of Africa.

Libyē, Inner (or Desert) [Map 7, 1.8, 8.1]: region (4.6) encompassing the Saharan regions south of the Roman provinces of Mauretania, Africa, and Cyrenaica and north of *Aithiopia. Since *Garamē is within Inner Libyē, it is probably to this that Ptolemy alludes when he writes that Septimius Flaccus "made a campaign out of Libyē."

Lilybaeum [Map 5, 1.12]: mod. Marsala, on the extreme western cape of Sicily.

Limyrikē [Map 8, 1.7]: roughly, the Malabar coast of southwest India.

Londinium [Map 4, 1.15]: on the site of the modern City of London.

Maiōtis, Lake [Map 6, 1.8, 2.1, 7.5]: mod. Sea of Azov.

Mēdia [Map 8, 1.12]: district (satrapy) of Persia, east of *Assyria and south and southwest of the Caspian Sea.

Mediterranean ("Our Sea") [Maps 4–6, 2.1, 7.5]

Meroē [Map 7, 1.7, 9–10, 15]: city (and surrounding district) near modern Kabushiya, Sudan, south of the junction of the *Nile and Atbara Rivers. This situation was misunderstood by Ptolemy (4.7.20–22), who describes the country of Meroē as a large island in the Nile, with the Nile properly so called on its west and the "Astabora" (Atbara) on its east.

Meroē, parallel through [1.23–24, 7.5–6]: 16$\frac{5}{12}$° N, along which the longest day is 13 equinoctial hours.

Mesopotamia [Map 8, 1.12]: on Ptolemy's map, this name is used only for the district between the Euphrates and Tigris rivers, which unite before flowing into the Persian Gulf. *Assyria to the northeast of the Tigris, and Babylonia southwest of the Euphrates, are separate districts.

Mysia (or Moesia) [Map 6, 1.16]: two Roman provinces south of the Danube and north of the main Balkan range, which separated them from Macedonia and *Thrace. Upper Mysia was the western province, corresponding to parts of modern Serbia and northwest Bulgaria.

Nile [Map 7, 1.15, 2.1].

Nile, lakes at source [Map 7, 1.9, 15, 17]: reports of these lakes, brought back by traders sailing down the east coast of Africa as far as *Rhapta, presumably refer to the great lakes of east central Africa, which include the actual sources of the White Nile (Lakes Victoria and Albert)—although it is doubtful whether the connection drawn between these lakes and the Nile was more than a guess.

Noricum [Map 5, 1.16]: Roman province in the eastern Alps between *Rhaetia and *Pannonia, roughly corresponding to the eastern part of mod. Austria.

Noviomagus [Map 4, 1.15]: mod. Chichester.

Ocean [Maps 3–4 and 6, 7.5, 7, 8.1]: the body of water outside the Straits of *Hēraklēs, and not believed to be enclosed by land. In Ptolemy's map the Ocean forms the west coasts of *Libyē and *Europe only. Parts of the Ocean are named by qualifying adjectives, including the Western Ocean (along the western edge of the map), the Duēcalidonian Ocean (north of *Albiōn), and the Sarmatian Ocean (north of *Sarmatia).

Okēlis, peninsula and trading post [Map 7, 1.7, 15]: the "straits" are the Bāb al-Mandab, which is the narrowest point of the Red Sea in the south. The "city" of *Dērē is on a cape (Ras Siyan?) on the African side, and Okēlis on the Arabian. In 6.7.7 Ptolemy calls the peninsula at Okēlis (Ras Bāb al-Mandab) "Cape Backward-run" (*Palindromos*). See Casson 1989, 116 and 157–158.

Opōnē, trading station [Map 7, 1.17]: probably mod. Xaafuun, Somalia. See Casson 1989, 132.

Pachynus [Map 5, 1.12, 15]: mod. Pachino, Sicily.

Palimbothra [Map 8, 1.12, 17]: a Greek rendering of Pataliputra, near mod. Patna, India.

Paloura [Map 8, 1.13]: in 7.1.15–16 Ptolemy places Paloura at the beginning of the Bay of *Ganges, just after "the setting-out point of those who sail to Chrysē" (referring to the district of that name north of *Sada, cf. 7.2.17). Perhaps Paloura was the modern Machilipatnam (or Masulipatam); see Casson 1989, 232.

Pannonia [Map 5, 1.16]: Roman province south of the Danube, and east of *Noricum, now mostly in Hungary.

Panōn, town [Map 7, 1.17]: somewhere between *Arōmata (Cape Guardafui) and *Opōnē on the coast of Somalia.

Paphos [Map 6, 1.15] (Cyprus).

Parthia [Map 8, 1.12]: district (satrapy) of Persia, east of *Media and southeast of the Caspian Sea.

Peloponnese [Map 6, 7.5] (Greece).

Persian Bay [Map 8, 7.5]: mod. Persian Gulf.

Phalangis mountain [Map 7, 1.17]: location uncertain, somewhere along the coast of Somalia between *Zingis and the *Bluff.

Pisae [Map 5, 1.15]: mod. Pisa, Italy.

Pontos, Sea of [Map 6, 1.16, 7.5]: mod. Black Sea.

Pontos, parallel through middle of [1.16]: 45° N, along which the longest day is 15½ equinoctial hours.

Prason, Cape [Map 7, 1.7–10, 14, 17, 2.1]: perhaps Cape Delgado (10°41' S, 40°38' E). The name probably alludes to a kind of seaweed (*prason*) that was imagined to infest the "Seaweedy" (Prasōdes) Sea extending from Menouthias Island (off Cape Prason) straight east across the Sea of *India to the sea south of the *Great Bay (7.2.1). See Desanges 1978, 332–333.

Propontis [Map 6, 7.5]: mod. Sea of Marmara.

Ptolemais Thērōn (i.e., Ptolemais of the Hunts) [Map 7, 1.8, 15]: station on the Sudanese coast of the Red Sea, established in the third century B.C. by

Ptolemy Philadelphos as a base for elephant-hunting expeditions. The site (probably between 18° and 19° N) is uncertain. See Casson 1989, 100–101.

Pyrenees [Map 5, 1.15].

Ravenna [Map 5, 1.15] (Italy).

Rhaetia (or Raetia) [Map 5, 1.16]: Roman Alpine province west of *Noricum, now divided between southern Germany, Austria, and Switzerland.

Rhapta, city and cape (also called Cape Rhapton) [Map 7, 1.9, 14, 17, 23]: the southernmost trading station known to Ptolemy along the east coast of Africa, named "sewn things" (*rhapta*), according to *Periplus* 16, for the boats stitched together there. Rhapta was somewhere on the coast of modern Tanzania within a degree or so either way of Dar es Salaam (6°49' N). The river that Ptolemy situates nearby may have been the Pangani or the Rufiji. See Casson 1989, 141.

Rhaptos, river [1.17]: see *Rhapta.

Rhodes [Map 6, 1.11–12] (Greece).

Rhodes, parallel through [1.12, 14, 20–21, 23–24, 7.5]: 36° N, along which the longest day is 14½ equinoctial hours.

Sachalitēs, Bay of [Map 8, 1.17]: the bay to the east of Cape *Syagros (Ra's Fartak), which Ptolemy calls the "Bay of Sachalitēs," is mod. Qamar Bay. The coast to the west of Ra's Fartak has no clearly defined bay, but only a gentle curvature. Nevertheless, Marinos' view of the relative locations of the Bay of Sachalitēs and Cape Syagros finds support in the *Periplus* 29–32, which places the Bay of Sachalitēs to the west of Cape Syagros and names the bay to the east "Omana." See Casson 1989, 165-166.

Sachalitēs, country [Map 8, 1.17]: part of *Arabia Felix along the Bay of *Sachalitēs (6.7.11).

Sacred Cape of Spain [Map 5, 1.12]: mod. Cape S. Vicente (Portugal).

Sada [Map 8, 1.13]: city in *India Beyond the Ganges, probably on the coast of Burma near mod. Sandoway.

Sakai [Map 8, 1.16]: nomadic inhabitants of the region about the Pamirs north of India (see *Kōmēdai).

Samos [Map 6, 1.7] (Greece).

Sarapiōn, anchorage of [Map 7, 1.17]: identified as Warshiikh (Somalia) by Casson (1989, 138–139).

Sardinia [Map 5, 1.12, 7.5] (Italy).

Sarmatia [Maps 3 and 7, 7.5]: a large region in the north of Ptolemy's Europe and Asia, east of Germania and north of the Sea of *Pontos and part of the *Hyrkanian Sea.

Sarmatian Ocean [7.5]: see *Ocean.

Sarmatians [1.8]: see *Sarmatia.

Satala [Map 6, 1.15]: mod. Sadagh, Turkey.

Schoinoi, Thirty: see *Thirty Schoinoi.

Sebennytos [Map 6, 1.15]: town (mod. Samannūd) after which one of the ancient mouths of the *Nile was named.

Sēra [Map 8, 1.11–12]: the "metropolis" of the *Sēres. Often identified as Luoyang, the Han capital of China in Ptolemy's time.

Sēres [1.11, 17]: the "Silk People," i.e., the Chinese as known by the inland trade routes, inhabiting the northeasternmost part of Ptolemy's *oikoumenē* (*Sērikē).

Sērikē [Map 8, 7.5]: the country of the *Sēres.

Sicily [Map 5, 1.12, 7.5] (Italy).

Simylla (or Timoula) [Map 8, 1.17]: trading station in *Limyrikē. The *Periplus* 53 also has "Semylla" and, like Ptolemy, places it east of the *Indus, near mod. Bombay.

Sinai [Map 8, 1.11, 17, 7.5]: the Chinese, as known from the commercial routes of the Indian Ocean by way of India. The name probably reflects the Ch'in (or Ts'in) Dynasty of the third century B.C. Ptolemy situates the country of the Sinai on the Sea of *India, east of *India Beyond the Ganges and south of the country of the *Sēres (who were in fact the same people, known through a different channel). The coast comprises half of the *Great Bay, the Bay of Beasts (*Thēriōdes*), and the Bay of *Sinai, with a pronounced southward trend (see *Zabai).

Sinai, Bay of the [Map 8, 1.13]: a large bay in the country of the *Sinai, north of *Kattigara. Sometimes identified with the Gulf of Tongking (but see *Great Bay). The dwellers about the bay are described (7.3.1) as "fish-eating *Aithiopians."

Sinai, metropolis of the [Map 8, 1.14, 17, 7.5]: in the geographical catalogue (7.3.6 and 8.27.12) this city is given the name Thinai (Theinai in some manuscripts); Ptolemy situates it a little to the north and east of *Kattigara. The *Periplus* (ch. 64), however, states that “Thina” is a city in the far (arctic!) north beyond India, and from it goods were transported overland to India. Thinai might be another reflection of Luoyang (see *Sēra).

Skythia [Map 8, 7.5, 8.1]: vast region of *Asia, running across the north edge of the *oikoumenē* from *Sarmatia on the west to the eastern edge of the map.

Skythians [1.8]: see *Skythia.

Smyrna, parallel through [1.12]: $38\frac{7}{12}$° N, along which the longest day is $14\frac{3}{4}$ equinoctial hours.

Soēnē (or Syēnē) [Map 7, 1.7, 9]: the modern Aswān. Soēnē was assumed by Greek geographers to lie exactly on the Tropic of Cancer.

Soēnē, parallel through [1.23–24, 7.5–7]: $23\frac{5}{6}$° N, through which the longest day is $13\frac{1}{2}$ equinoctial hours.

Sogdians [Map 8, 1.16]: inhabitants of the region north of *Baktria between the rivers Oxus and Iaxartes, around the mod. Samarkand and Bukhara (Uzbekistan).

Stone Tower [Map 8, 1.11–12, 17]: apparently a trading station of the Silk Route in the Pamirs (see *Kōmēdai). There is more than one place in this district now called Tash Kurghan (“Stone Tower”), but it is uncertain which, if any, of these was Ptolemy’s Stone Tower.

Strymōn [Map 6, 1.15]: mod. river Strimon (Struma) in northern Greece and Bulgaria.

Syagros, Cape [Map 8, 1.17]: Ra’s Fartak, on the south coast of the Arabian peninsula. See *Sachalitēs.

Tainaros [Map 6, 1.12]: Cape Matapan (or Tainaron), on the south coast of the Peloponnese.

Tamala [Map 8, 1.13]: city on the coast of *India Beyond the Ganges, probably near the mouth of the Irrawaddy River, Burma.

Tanais, River [2.1, 7.5]: mod. Don River. Its mouth is at the northern end of Lake *Maiōtis.

Taprobanē [Map 8, 1.14, 7.5, 8.1]: mod. Sri Lanka.

Tarraco [Map 5, 1.15]: mod. Tarragona, Spain.

Tergestē [Map 5, 1.15]: mod. Trieste, Italy.

Thēna (*Theainai*) [Map 5, 1.15]: on the coast of the Roman province of Africa, near mod. Sfax, Tunisia.

Thirty Schoinoi [Map 7, 1.9]: As Ptolemy states in 4.5.74 and 4.7.32, the district along the Nile south from *Soēnē to Pselkis (mod. El Dakka, now beneath Lake Nasser) was called the "Twelve Schoinoi," and the adjacent district further south and on the west side of the river was called the "Thirty Schoinoi." On the *schoinos* as a unit of distance see p. 14.

Thrace [Map 6, 1.15–16]: district east of Macedonia, now the European part of Turkey and parts of northeastern Greece and southeastern Bulgaria.

Thulē [Map 4, 1.7]: originally the name of a possibly fictitious island close to the Arctic Circle described by Pytheas of Massalia (late fourth century B.C.), a Greek traveler whose veracity was questioned in antiquity. By Ptolemy's time Thulē probably meant the mod. Shetland Islands.

Thulē, parallel through [1.20–21, 23–24, 7.5–6]: 63° N, along which the longest day is 20 equinoctial hours.

Tigris, river [Map 8, 1.12].

Tileventus, river [Map 5, 1.15]: mod. river Tagliamento (northeastern Italy).

Timoula [1.17]: See *Simylla.

Toniki [Map 7, 1.17]: perhaps the place referred to by the *Periplus* (ch. 15) as the "run of Nikōn," immediately following that of *Sarapiōn, and presumably at about latitude 2°. See Casson 1989, 134.

Trapezous [Map 7, 1.15]: mod. Trabzon, Turkey.

Trōglodytikē [1.8–9]: the manuscripts of the *Geography* (as well as some other classical sources) give the name of part of the African coast as "Trōglodytikē," a name that would mean "the country of dwellers in holes." This seems to be the product of a false etymology, the correct name being "Trōgodytikē" (see *LSJ* s.v. Τρωγοδύται), but the mistake may have been Ptolemy's. In 4.7.27 Ptolemy applies this name to the east coast of *Aithiopia as far south as Elephant Mountain, which is just north of *Arōmata. *Ptolemais Thērōn lies within this stretch of coast. However, the coast that Diogenes was driven along (1.9) was *south* of Arōmata, so that Diogenes (or Marinos) must have used the term in a broader sense, taking in what Ptolemy calls *Barbaria or *Azania.

Unknown land [Map 7, 1.17, 2.1, 7.5]: the *Ocean accounts for only part of the western and northern sides of the quadrilateral framing Ptolemy's *oikoumenē*. Where various regions of *Europe and *Asia (*Sarmatia, *Skythia, *Sērikē, the country of the *Sinai) reach the north and east sides, and *Aithiopia reaches the west and south sides, Ptolemy must hypothesize "unknown lands" that fall outside the world map. The Sea of *India is also supposed by Ptolemy to be closed off to the south by "unknown land" linking Aithiopia with the country of the Sinai. The coast of this is left undefined in the description of Aithiopia in 4.8, which stops at Cape *Prason. In the description of the country of the Sinai (7.3), the last place assigned a position is *Kattigara, and at the end of the chapter Ptolemy writes, "After Kattigara, [the country] is bounded to the west by unknown land that surrounds the Seaweedy Sea [for which see Cape *Prason] as far as Cape Prason, where, as has been said, begins the bay of the Rough [*Tracheia*] Sea, linking the land to Cape Rhapton and the southern parts of Azania." Ptolemy gives no instructions for the drawing of this coastline (it is, after all, *unknown*) and leaves open the possibility that it lies below the southern edge of the map, at 16⁵⁄₁₂° S, a few degrees south of Kattigara and Cape Prason.[1] However, the caption for the world map (7.5) refers to the enclosure of the Sea of India more than once, and the caption for the picture of the ringed globe (7.7) expressly states that the drawing portrayed the sea as surrounded by land.

Western Ocean [7.5, 8.1]: see *Ocean.

Zabai [Map 8, 1.14]: city near the cape that marks the beginning of the *Great Bay.

Zingis, Cape [Map 7, 1.17]: unidentified cape south of *Opōnē.

[1]It is surely for this reason, and not because Ptolemy changed his mind about the Sea of India, that 2.1 mentions only the isthmus at *Hērōopolis as a place where *Libyē and Asia join. There he is defining the limits of the three continents solely for the purposes of the catalogue.

# *Bibliography*


Andersen, K. 1987. "The Central Projection in One of Ptolemy's Map Constructions." *Centaurus* 30, 106–113.

Aujac, G. 1993. *Claude Ptolémée astronome, astrologue, géographe. Connaissance et représentation du monde habité.* Paris.

Bagrow, L. 1943. "The Origin of Ptolemy's Geographia." *Geografiska Annaler* 27, 318–387.

Barbier de Meynard, C., and Pavet de Courteille, eds. and trans. 1861–1877. *Maçoudi: Les prairies d'or.* 9 vols. Paris.

Berger, H. 1903. *Geschichte der wissenschaftlichen Erdkunde der Griechen.* 2nd ed. Leipzig.

Berggren, J. L. 1991. "Ptolemy's Maps of Earth and the Heavens: A New Interpretation." *Archive for History of Exact Sciences* 43, 133–144.

Bosio, L. 1983. *La Tabula Peutingeriana: Una descrizione pittorica del mondo antico.* I Monumenti dell'Arte Classica 2. Rimini.

Carra de Vaux, B., trans. 1897. *Le livre de l'avertissement et de la revision [par] Maçoudi.* Paris.

Casson, L. 1989. *The Periplus Maris Erythraei.* Princeton.

Codazzi, A. 1948–1949. *Le edizioni quattrocentesche e cinquecentesche della Geografia di Tolomeo.* Milano.

Cuntz, O. 1923. *Die Geographie des Ptolemaeus: Galliae Germania Raetia Noricum Pannoniae Illyricum Italia.* Berlin.

———. 1929. *Itineraria Romana.* Vol. 1. Leipzig.

Desanges, J. 1964. "Les territoires gétules de Juba II." *Revue des Études Anciennes* 66, 33–47.

———. 1978. *Recherches sur l'activité des Méditerranéens aux confins de l'Afrique (VI*[e] *siècle avant J.-C.–IV*[e] *siècle après J.-C.).* Rome.

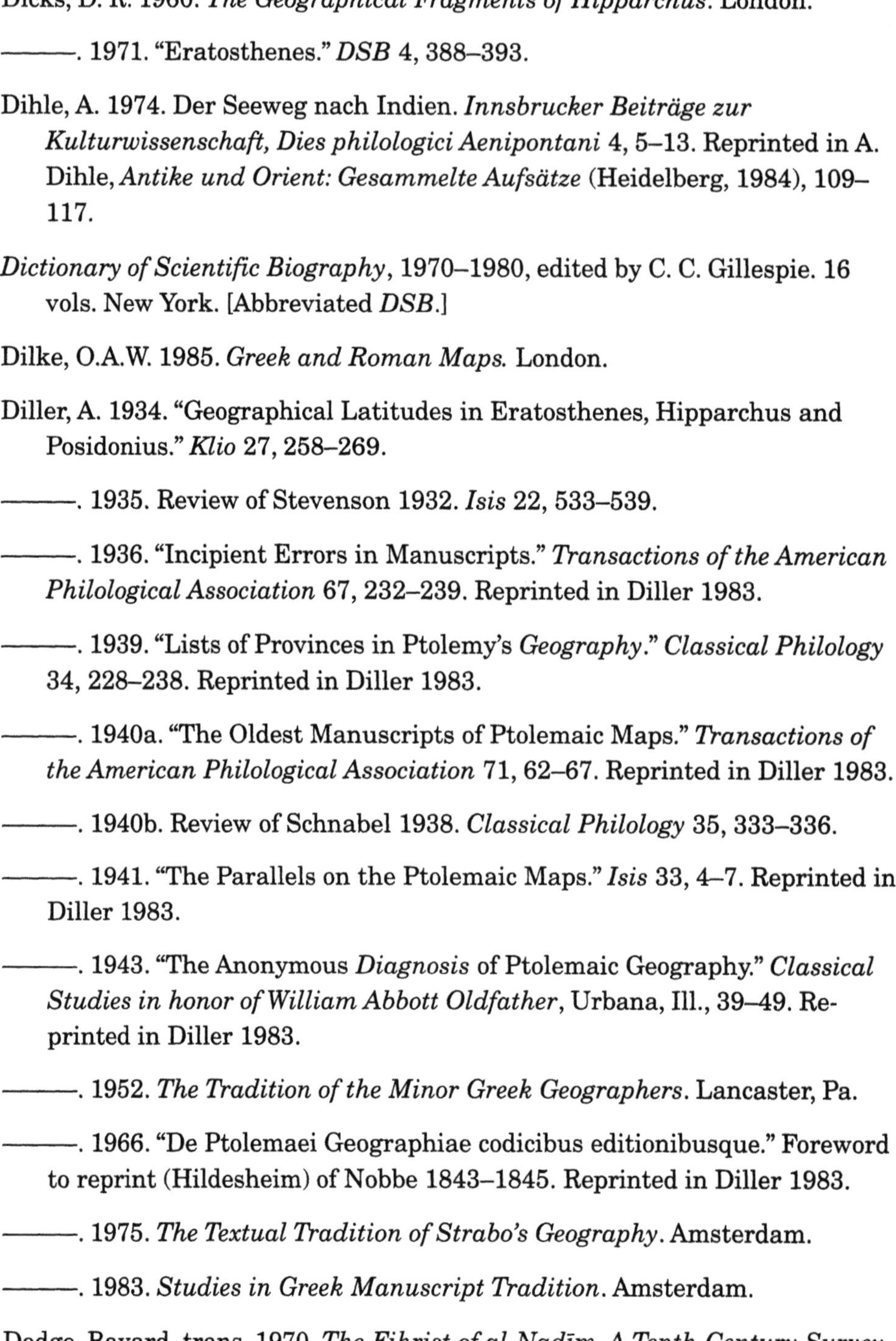


Dicks, D. R. 1960. *The Geographical Fragments of Hipparchus*. London.

———. 1971. "Eratosthenes." *DSB* 4, 388–393.

Dihle, A. 1974. Der Seeweg nach Indien. *Innsbrucker Beiträge zur Kulturwissenschaft, Dies philologici Aenipontani* 4, 5–13. Reprinted in A. Dihle, *Antike und Orient: Gesammelte Aufsätze* (Heidelberg, 1984), 109–117.

*Dictionary of Scientific Biography*, 1970–1980, edited by C. C. Gillespie. 16 vols. New York. [Abbreviated *DSB*.]

Dilke, O.A.W. 1985. *Greek and Roman Maps.* London.

Diller, A. 1934. "Geographical Latitudes in Eratosthenes, Hipparchus and Posidonius." *Klio* 27, 258–269.

———. 1935. Review of Stevenson 1932. *Isis* 22, 533–539.

———. 1936. "Incipient Errors in Manuscripts." *Transactions of the American Philological Association* 67, 232–239. Reprinted in Diller 1983.

———. 1939. "Lists of Provinces in Ptolemy's *Geography*." *Classical Philology* 34, 228–238. Reprinted in Diller 1983.

———. 1940a. "The Oldest Manuscripts of Ptolemaic Maps." *Transactions of the American Philological Association* 71, 62–67. Reprinted in Diller 1983.

———. 1940b. Review of Schnabel 1938. *Classical Philology* 35, 333–336.

———. 1941. "The Parallels on the Ptolemaic Maps." *Isis* 33, 4–7. Reprinted in Diller 1983.

———. 1943. "The Anonymous *Diagnosis* of Ptolemaic Geography." *Classical Studies in honor of William Abbott Oldfather*, Urbana, Ill., 39–49. Reprinted in Diller 1983.

———. 1952. *The Tradition of the Minor Greek Geographers*. Lancaster, Pa.

———. 1966. "De Ptolemaei Geographiae codicibus editionibusque." Foreword to reprint (Hildesheim) of Nobbe 1843–1845. Reprinted in Diller 1983.

———. 1975. *The Textual Tradition of Strabo's Geography*. Amsterdam.

———. 1983. *Studies in Greek Manuscript Tradition*. Amsterdam.

Dodge, Bayard, trans. 1970. *The Fihrist of al-Nadīm. A Tenth-Century Survey of Muslim Culture.* 2 vols. New York.

Engels, D. 1985. "The Length of Eratosthenes' Stade." *American Journal of Philology* 106, 298–311.

Fischer, J. 1932a. *De Cl. Ptolemaei vita operibus geographia praesertim eiusque fatis.* 2 vols. Codices e Vaticanis selecti 18. Vatican City.

———. 1932b. *Claudii Ptolemaei Geographiae codex Urbinas graecus 82 phototypice depictus*. 2 vols. Codices e Vaticanis selecti 19. Vatican City.

Furneaux, H. 1896. *P. Cornelii Taciti Annalium ab excessu divi Augusti libri: The Annals of Tacitus*. 2nd ed. 2 vols. Oxford.

Gallazzi, C., and B. Kramer. 1998. "Artemidor im Zeichensaal. Eine Papyrusrolle mit Text, Landkarte und Skizzenbüchern aus späthellenistischer Zeit." *Archiv für Papyrusforschung* 44, 189–208 and plate xxi.

Ginzel, F. K. 1899. *Spezieller Kanon der Sonnen- und Mondfinsternisse.* Berlin.

Goldstein, B. R. 1967. *The Arabic Version of Ptolemy's* Planetary Hypotheses. *Transactions of the American Philosophical Society* 57.4. Philadelphia.

Halma, N. 1822–1825. Vol. 1: *Commentaire de Théon d'Alexandrie sur le livre III de l'Almageste de Ptolemée; Tables manuelles des mouvemens des astres.* Paris, 1822. Vols. 2 and 3: *Tables manuelles astronomiques de Ptolemée et de Théon.* Paris, 1823 and 1825.

Hamilton, N. T., N. M. Swerdlow, and G. J. Toomer. 1987. "The Canobic Inscription: Ptolemy's Earliest Work." In *From Ancient Omens to Statistical Mechanics: Essays on the Exact Sciences Presented to Asger Aaboe*, edited by J. L. Berggren and B. R. Goldstein, 55–73. Copenhagen.

Heiberg, J. L. 1907. *Claudii Ptolemaei opera quae exstant omnia.* Vol. 2, *Opera astronomica minora.* Leipzig.

Hewsen, R. 1971. "The *Geography* of Pappus of Alexandria: A Translation of the Armenian Fragments." *Isis* 62, 187–207.

———. 1992. *The Geography of Ananias of Širak*. Wiesbaden.

Honigmann, E. 1929. *Die sieben Klimata und die* πόλεις ἐπίσημοι. Heidelberg.

———. 1930. "Marinos von Tyros." *RE* 28, 1767–1796.

Kennedy, E. S. 1987. *Geographical Coordinates of Localities from Islamic Sources.* Frankfurt.

Kugéas, S. 1909. "Analekta Planudea." *Byzantinische Zeitschrift* 18, 106–146.

Lejeune, A. 1948. *Euclide et Ptolémée, deux stades de l'optique géométrique greque*. Louvain.

Miller, K. 1916. *Die Peutingersche Tafel*. In K. Miller, *Mappaemundi: Die älteste Weltkarten*. 6 vols. Stuttgart. Reprinted separately, Stuttgart, 1962.

Mommsen, T. 1881. "Ammians Geographica." *Hermes* 16, 602–636.

Müller, C. 1855–1861. *Geographi Graeci minores*. 2 vols. Paris.

Müller, C., with C. T. Fischer. 1883–1901. *Claudii Ptolemaei Geographia*. 2 vols. Paris.

Mžik, H. 1916. *Afrika nach der arabischen Bearbeitung der* Γεωγραφικὴ ὑφήγησις *des Claudius Ptolemaeus von Muḥammad ibn Mūsā al-Ḫuwārizmī*. Kaiserliche Akademie der Wissenschaften in Wien, Philosophisch-historische Klasse, Denkschriften 59.4. Wien.

———. 1926. *Das Kitāb Ṣūrat al-Arḍ des Abū Ǧa'far Muḥammad ibn Mūsā al–Ḫuwārizmī*. Leipzig.

Mžik, H., with F. Hopfner. 1938. *Des Klaudios Ptolemaios Einführung in die darstellende Erdkunde. Erster Teil. Theorie und Grundlagen der Darstellenden erdkunde*. Wien.

Neugebauer, O. 1938–1939. "Über eine Methode zur Distanzbestimmung Alexandria-Rom bei Heron." *Kgl. Danske Vidensk. selsk., Hist.-filol. Medd.* 26.2 and 26.7.

———. 1959. "Ptolemy's *Geography*, Book VII, Chapters 6 and 7." *Isis* 50, 22–29. Reprinted in Neugebauer 1983.

———. 1975a. *A History of Ancient Mathematical Astronomy*. 3 vols. New York.

———. 1975b. "A Greek World Map." *Le Monde Grec, Hommages à Claire Préaux*, 312–317. Bruxelles. Reprinted in Neugebauer 1983.

———. 1983. *Astronomy and History. Selected Essays*. New York.

Nobbe, C.F.A., ed. 1843–1845. *Claudii Ptolemaei Geographia*. 3 vols. Leipzig. (Reprint, Hildesheim, 1966.)

Nordenskiöld, A. E. 1889. *Facsimile-Atlas to the Early History of Cartography with Reproductions of the Most Important Maps Printed in the XV and XVI Centuries*. Trans. J. A. Ekelöf and C. R. Markham. Stockholm.

Polaschek E. 1959. "Ptolemy's *Geography* in a New Light." *Imago Mundi* 14, 17–37.

———. 1965. "Klaudios Ptolemaios, Das geographische Werk." *RE* suppl. 10, 680–833.

Pothecary, S. 1995. "Strabo, Polybios, and the Stade." *Phoenix* 49, 49–67.

*RE* = Pauly, A., G. Wissowa, and W. Kroll, eds. *Paulys Realencyclopädie der classischen Altertumswissenschaft*. 2nd ed. München, 1893–.

Renou, L. 1925. *La géographie de Ptolémée. L'Inde (VII, 1–4)*. Paris.

Rome, A. 1927. "L'Astrolabe et le Météoroscope d'après le commentaire de Pappus sur le 5e livre de l'Almageste." *Annales de la Société Scientifique de Bruxelles* 47, Deuxième partie, mémoires, 77–102.

Sachs, A., and H. Hunger. 1988–. *Astronomical Diaries and Related Texts from Babylonia*. Österreichische Akademie der Wissenschaften, Phil.–hist. Klasse, Denkschriften 195, 210, 246 (more volumes forthcoming). Wien.

Schnabel, P. 1930. "Die Entstehungsgeschichte des kartographischen Erdbildes des Klaudios Ptolemaios." *S.B. d. Preussischen Akademie der Wissenschaften, phil.–hist. Klasse* 14, 214–250.

———. 1938. *Text und Karten des Ptolemäus*. Quellen und Forschungen zur Geschichte der Geographie und Völkerkunde 2. Leipzig.

Schoff, W. H. 1914. *The Parthian Stations of Isidore of Charax*. Philadelphia.

Smith, A. M. 1996. *Ptolemy's Theory of Visual Perception: An English Translation of the Optics with Introduction and Commentary.* Transactions of the American Philosophical Society 86.2. Philadelphia.

Standish, J. F. 1970. "The Caspian Gates." *Greece & Rome*, n.s. 17, 17–24.

Stein, A. 1928. *Innermost Asia. Detailed Report of Explorations in Central Asia, Kan-su and Eastern Iran, Carried Out and Described under the Orders of H.M. Indian Government.* 4 vols. Oxford.

Stevenson, E. L. 1932. *Geography of Claudius Ptolemy*. New York. (Reprint, New York, 1991.)

Stückelberger, A. 1996. "Planudes und die Geographia des Ptolemaios." *Museum Helveticum* 53, 197–205.

Syme, R. 1958. *Tacitus*. 2 vols. Oxford.

Taisbak, C. M. 1974. "Posidonius Vindicated at All Costs? Modern Scholarship Versus the Stoic Earth Measurer." *Centaurus* 18, 253–269.

Toomer, G. J. 1975. "Ptolemy." *DSB* 11, 186–206.

———. 1978. "Hipparchus." *DSB* 15, 207–224.

———. 1984. *Ptolemy's Almagest*. London.

Turyn, A. 1964. *Codices graeci vaticani saeculis XIII et XIV scripti annorumque notis instructi*. Codices e Vaticanis selecti 28. Vatican City.

Wieber, R. 1974. *Nordwesteuropa nach der arabischen Bearbeitung der Ptolemäischen Geographie von Muḥammad B. Mūsā Al-Ḫwārizmī*. Walldorf–Hessen.

———. 1995. "Marinos von Tyros in der Arabischen Überlieferung." In M. Weinmann-Walser, ed., *Historische Interpretationen Gerold Walser zum 75. Geburtstag dargebracht von Freunden, Kollegen und Schülern*, 161–190. Stuttgart.

Wilberg, F. W., with C.H.F. Grashof. 1838–1845. *Claudii Ptolemaei geographiae libri octo*. 6 fascicles. Essen.

Wilson, N. G. 1981. "Miscellanea Palaeographica." *Greek, Roman, and Byzantine Studies* 22, 395–404.

———. 1983. *Scholars of Byzantium*. London.

# *Index*

*For references to geographical localities in Ptolemy's text, see Appendix H.*